ULTRASONIC MICRO/NANO MANIPULATIONS

Principles and Examples

ULTRASONIC MICRO/NANO MANIPULATIONS

Principles and Examples

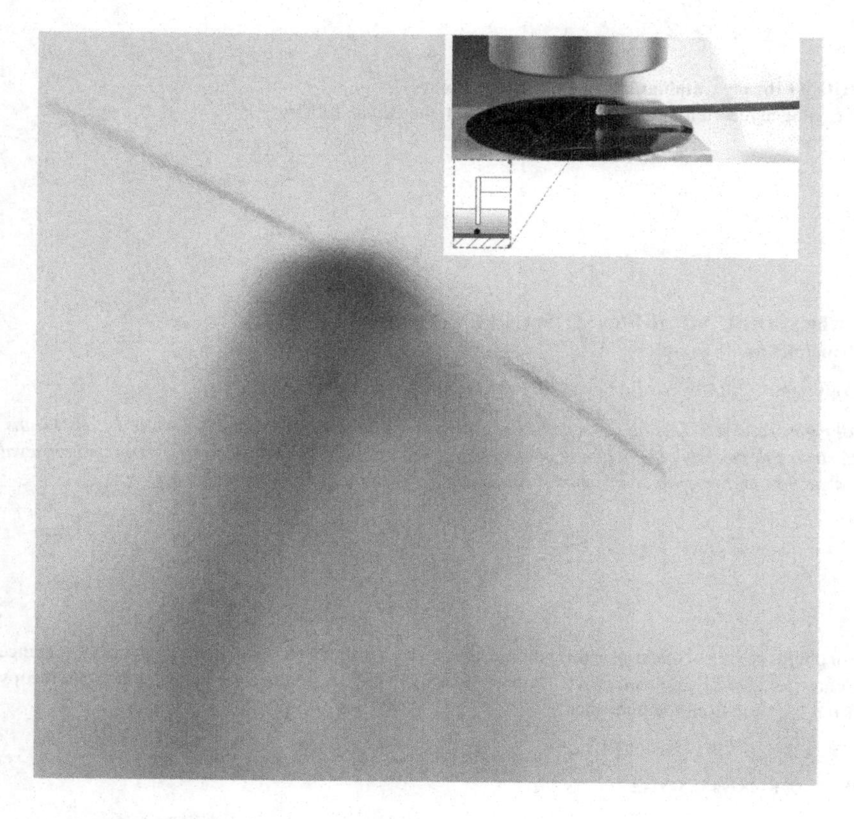

Junhui Hu

Nanjing University of Aeronautics and Astronautics, China

NEW JERSEY • LONDON • SINGAPORE • BEIJING • SHANGHAI • HONG KONG • TAIPEI • CHENNAI

Published by

World Scientific Publishing Co. Pte. Ltd.
5 Toh Tuck Link, Singapore 596224
USA office: 27 Warren Street, Suite 401-402, Hackensack, NJ 07601
UK office: 57 Shelton Street, Covent Garden, London WC2H 9HE

British Library Cataloguing-in-Publication Data
A catalogue record for this book is available from the British Library.

ULTRASONIC MICRO/NANO MANIPULATIONS
Principles and Examples

ISBN 978-981-4525-31-2

Printed in Singapore

Preface

This book is designed for the scientists, engineers, students and research project managers who are engaging in the research and development in ultrasonic manipulation technology or are interested in this technology. It gives the basic physical principles of ultrasonic micro/nano manipulations, and detailed methods of implementing these principles. Lots of examples are given in this book, to help the readers better understand the applications of these principles and characteristics of ultrasonic manipulators utilizing these principles.

Demands for high-performance micro/nano manipulations, which come from the manufacture of microelectronic and photonic devices, biomedical apparatus, nanoscience and nanotechnology, renewable energy, environment protection, high-end appliances, etc. have been rapidly increasing in recent years. However, there are very few books on ultrasonic manipulation technology, which is one of the important means in micro/nano manipulations. I hope that this book would make a contribution to the development and application of micro/nano manipulation technology.

This book consists of seven chapters. Chapter 1 is on the diversity of required actuation functions, and generalities of ultrasonic micro/nano manipulations. Chapter 2 is on the physics in ultrasonic micro/nano manipulations whereas Chapters 3-6 give examples of how to implement these principles and characteristics of the devices. In Chapter 3, examples of ultrasonic contact and noncontact type trapping of micro solids are given and discussed. In Chapter 4, examples of ultrasonic extraction, driving and removal of micro solid are given and discussed. In Chapter 5, examples of ultrasonic manipulations of nanoscale entities are given and discussed. In Chapter 6, examples of ultrasonic microfluidic manipulations are given and discussed. The examples given in Chapters 3-6 come from the work which my research group did in the past 10 years. In Chapter 7, I give concluding remarks for the principles and examples listed in this book, which includes the position of ultrasonic micro/nano manipulation in ultrasonic technology, manipulation functions provided by this technology, features and limitations of this technology, and big academic and technological challenges which we are facing in the development of this technology.

Although this book involves the MEMS based acoustic manipulation little, the physical principles, demonstrated in this book, can also be applied in the MEMS based acoustic manipulation.

Many outstanding scientists contributed greatly to the physical principles of ultrasonic micro/nano manipulations, design and fabrication of the devices based on these physical principles, and modeling, characterization and optimization of these devices. Due to the limited topic number in this book, only the work by some of them, which is relevant to the topics, is cited. However, the work of the others is equally important to micro/nano manipulation technology.

I would like to thank my students for their contributions and efforts to the research work described in the examples. They are Mr. Armand Kertoputro Santoso, Dr. Yanyan Liu, Dr. Satyanarayan Bhuyan, Dr. Jinlong Du, Dr. Jianbo Yang, Ms. Jun Xu, Ms. Chiaolin Tay, Ms. Yanmin Cai, Mr. Chinlee Tan, Mr. Wenyao Hu, Ms. Libin Ong, Mr. Changhan Yeo, Ms. Xueyi. Zhang, Mr. Y. Zheng, Mr. Zhi Wen Tan, Ms. Su Gui Gisela Teo, Mr. Ning Li, Mr. Yujie Zhou, Mr. Xiaobo Zhu, Mr. Huaqing Li, Mr. Xiaolong Lu, and Mr. Qi Zhang. Also, I should like to thank Prof. Yongxiao Chen, Prof. Yoshiro Tomikawa, Prof. Kentaro Nakamura, Prof. Tieying Zhou, Prof. Chunsheng Zhao and many other peers, for their kind support and invaluable advice in my academic career. Finally, I would like to express my gratitude to Dr. Hanmin Peng who also contributed to the production of this book, and my wife Qun Yue for her understanding and support.

Junhui Hu

Foreword

The book *Ultrasonic Micro/Nano Manipulations* appears just in time. It is filling a gap in the currently available review literature in this emerging subject area. It is dealing with interactions between ultrasonic fields and particles in suspensions, whereby the particles may be solid, soft, liquid, bubbles or even biological cells, while as dispersion medium water is primarily used. In the early 1990s the *particles in an ultrasonic field* research was founded by late Prof. Terence Coakley, Univ. Cardiff, UK, and was than around the year 2000 strongly promoted by a large and very successful European Marie Curie network entitled *Ultrasonic Separation of Suspended Particles* that was funded by the European Commission and was coordinated by the author of this foreword.

From the involved different groups several key researchers emerged like Stefan Radel, Jeremy Hawkes, Martyn Hill, which further pushed forward the development of acoustic bio-cell filters, analytical devices and miniaturization of acoustic-field/flow devices, and built the kernel of numerous dedicated special sessions at all major scientific congresses covering ultrasound applications. Currently, the most recognized contributions to acoustic-field/flow experiments come from Profs. Thomas Lemell, Lund University, and Martin Wiklund, KTH Stockholm, with focus on ultrasonic standing wave manipulation in combination with *Lab on a Chip* technology and micro-analytics, as well as from Prof. Duval, ETH Zürich, with emphasis on general micro-particle manipulation. However, while the mentioned key researchers merely utilize ultrasonic standing waves, for which acoustic streaming is rather an unwanted and disturbing effect, the author of the current work, Prof. Junhui Hu, is uniquely focusing on acoustic streaming as useful phenomenon for manipulating micro/nano particles.

This book deals with very diverse concepts of using ultrasound for micro/nano particle manipulation, but the treatment always follows the same structure and systematically demonstrates the principles and feasible device concepts, which can be used to realize those various particle manipulation tasks. Therefore the content is, although comprehensive, described in a compact form and helps readers to understand the potential applications of the principles and characteristics of ultrasonic manipulators. I am convinced that this volume will strongly promote the development and application of micro/nano manipulation technology and will trigger research and development towards many of the numerous fascinating potential applications. This book is indispensible for all students and scientists that are already working in this emerging field or are interested to join in. In this sense, I wish the author and his readers many future editions of this book and successful developments of new breath-taking particle manipulating devices, respectively.

Prof. Dr. Ewald Benes
Secretary General and former President and Chairman
of the International Congress on Ultrasonics
Vienna University of Technology, Institute of Applied Physics

Contents

Chapter 1

Introduction

In the first section of this chapter, the micro/nano manipulation functions, which are needed by the emerging industries and cannot be realized or efficiently realized by the conventional actuators, are given and described, and principles and basic features of the physical methods which are tried to realize the micro/nano manipulation functions are briefed. In the second section, functions and principles of ultrasonic micro/nano manipulations are given and described, and their basic features are summarized and discussed from the point of view of practical applications.

1.1 Diversity of Actuation

Conventional actuation mainly consists of rotary and linear driving, which can be realized by electromagnetic and ultrasonic motors. Electromagnetic and ultrasonic motors drive their rotors to rotate or push their sliders linearly. The conventional actuation is an important foundation of modern industries and supports our daily life. However, with the development of the fields such as micro/nano fabrication, nanoscience, biomedicine, renewable energy, etc., more and more actuation functions are being required, which include but not limited to:

(1) Trapping (or capture), orientation, transfer, release, sorting, revolution, spin, removal and concentration of solid and soft micro/nanoscale objects;
(2) Extension and winding of soft micro/nanoscale objects;
(3) Dispersion, agglomeration and concentration distribution control of micro/nanoscale objects in suspensions;

(4) Generation, transportation, merge, split and rotation of micro/nano liter droplets.

These actuation tasks are also called manipulations. Owing to the limited driving forms (rotary and linear) and operating principles of electromagnetic and ultrasonic motors, most of the above listed manipulations cannot be effectively and efficiently realized by them. For many scientists and engineers in micro/nano fabrication, biomedicine, nanoscience, renewable energy and other areas, it is a big challenge to effectively and efficiently realize these manipulations.

To fulfill the above listed actuation (or manipulation) functions, researchers in various academic areas have proposed and investigated different strategies. These strategies can be classified as optical, magnetic, electric, mechanic, microfludic and acoustic methods, based on the physical principles they use. They can also be classified as contact and noncontact methods, based on whether manipulated samples are in contact with the manipulators. The optical, magnetic, electrical, mechanical and microfluidic methods are briefed as follows.

In the optical method, various laser beams and optical fields are used to generate the optical radiation force, which is caused by the light momentum transfer upon light absorption or scattering by dielectric objects, to trap or/and rotate micro/nanoscale objects. Dr. Arthur Ashkin at AT & T Bell Lab first demonstrated the trapping of an individual microparticle in 1970 [1]. The optical method can trap and rotate individual micro/nanoscale entities with a wide density range in a noncontact means, and its trapping force can be up to the order of magnitude of piconewtons. Ashkin was able to trap particles with 10 to 10,000 nm diameter. Other scientists extended the technique to trap smaller particles. Steven Chu made use of the resonant laser and a magnetic gradient trap to trap neutral atoms with 0.1 nm diameter, which was used in the research earning him the 1997 Nobel Prize in Physics along with Claude Cohen-Tannoudji and William D. Phillips [2, 3]. The optical method could be harmful to biological samples owing to the heat generation in a laser beam [4,5].

In the magnetic method, magnetic field is used to manipulate magnetized micro/nanoscale samples or micro/nanoscale samples with magnetic components such as Ni caps [6,7]. It can align large amounts of micro/nanoscale samples if the samples contain the materials with high magnetic susceptibility. For practical applications, the main drawback of this method is that the manipulating force is small and the sample materials are limited.

In the electric method, dielectrophoresis, electroosmosis and electrothermal effects are used to trap micro/nanoscale objects [8-10]. Among the electric method, the dielectrophoresis effect is most commonly used, and has good application in cell sorting. It can separate cells with different dielectric properties. In the dielectrophoresis effect, spatial non-uniformity of electric field is used to generate a force on dielectrics in it, and the force is parallel to the electric field. The main drawback of the dielectrophoresis method is in the precise positioning of samples. The author's research group recently demonstrated a strategy to use hydrogen bubbles generated by electrolysis of water to manipulate microobjects in water on a substrate surface. Manipulation functions implemented by this strategy include direct rotary step-driving and attitude adjustment of a single microparticle, controlled release of a single sticky microparticle, etc. The device operating with this strategy has a very simple structure, little selectivity to material property of manipulated samples, and good controllability.

The mechanical method is usually based on micro grippers and the AFM probe assisted manipulations. Micro grippers, which are usually fabricated by MEMS process, can pick and place micro/nanoscale objects by electrostatic, piezoelectric, thermal and pneumatic actuations [8, 11, 12]. This method is reliable and stable in the pick-and-place manipulation. Demerits of this method include that the electric field or thermal gradient may be harmful to the samples, and manipulated sample may be contaminated owing to the contact between the sample and manipulator. A microscopic system assisted with two AFM probes can pick and place micro/nanoscale components to build 2D or 3D micro/nanoscale structures [13-15]. However, the systems are usually ineffective because they have to use bulky driving systems with sophisticated control functions and a series of imaging processing units.

In the microfluidic method, micro entities can be trapped, transferred, concentrated and diluted in the suspension flowing through microfluidic channels, and droplets can be merged and split [8, 16, 17]. The flow in microfluidic channels may be caused by the hydrodynamic force, capillary, gravity, wetting and adhesion, and nonlinear effects of sound. Combined with other manipulation methods such as the optical, electrical or acoustic method, it can have more powerful manipulating functions such as assembling nano entities on single molecules and trapping single molecules. Merits of this method include small sample consumption, portability, disposable use, no harmful effect on biological samples, etc. However, it is very difficult to use this method to implement some manipulations such as assembling micro/nanoscale structures.

The acoustic manipulation technology utilizes the physical effects of sound to manipulate micro/nanoscale solids [18-24] and micro/nanoliter droplets and bubbles [25-29]. It has the features such as no selectivity to the material properties of manipulated samples, little heat damage to manipulated samples (in some methods), diverse manipulation functions, simple and compact device structures, etc. Operating frequency in acoustic micro/nano manipulations may be in the ultrasonic range (>20 kHz) or several ten to several hundred Hz (in audible sound range) [28, 30]. The acoustic micro/nano manipulation with operating frequency in the ultrasonic range is called ultrasonic micro/nano manipulation. Most acoustic manipulation techniques make use of ultrasonic frequency.

1.2 Generalities of Ultrasonic Micro/Nano Manipulations

Functions provided by the ultrasonic micro/nano manipulation technology include the trapping, transfer, release, sorting, alignment, pumping, spin, revolution and removal of micro/nanoscale or larger solids in air and liquid, and the adsorption, extraction, generation, split, merge and rotation of micro/nanoliter droplets. In an ultrasonic manipulation device, the manipulation is implemented by ultrasonic field generated by a transducer in the device. Major sound fields employed by acoustic manipulation devices include the complex sound field in the vicinity of a radiation source, standing wave, leaky sound field, surface acoustic wave (SAW), focused sound beam and oscillating bubble induced sound waves.

Acoustic radiation force (or pressure) and acoustic streaming are two major physical effects used in ultrasonic manipulations. Acoustic radiation force may be used to manipulate micro objects, and acoustic streaming to manipulate nano objects. Acoustic radiation force may push multiple micro objects or an individual micro object to the nodal (or anti-nodal) position of an ultrasonic field, focal point of an ultrasonic beam, and radiation source surface. Acoustic streaming can not only disperse micro/nano objects, but also trap, align and rotate micro/nano objects.

The traveling wave around a nodal point of asymmetrically or non-uniformly excited plate may be used to drive micro objects to spin or revolve, and Chladni effect be used to push micro objects to the nodal point of a vibrating plate along a helical path or linearly. Unlike the methods based on acoustic radiation force and acoustic streaming, these two methods can manipulate the samples in vacuum. Sound induced decrease of intermolecular force may be used to implement many manipulation functions of droplets such as adsorption, merge, split, transport, generation, etc.

Based on whether a manipulated sample is in contact with the manipulating probe, the ultrasonic manipulation can be classified into contact and noncontact manipulations (see the examples in Sections 5.1 and 5.2). Based on the number of manipulated objects, the ultrasonic manipulation can be classified into single object manipulations and multiple object manipulations (see the examples in Sections 4.4 and 5.1-5.4). Based on manipulation scale, the ultrasonic manipulation can be classified into small- and large-scale manipulations. In the examples of this book, dust removal for photovoltaic panels is a large-scale manipulation (see Section 4.4), and the others are small-scale manipulations.

The main features of ultrasonic manipulations are listed and described as follows:

(1) No selectivity to the material properties of manipulated samples.

When the sound field for manipulation is proper, there is always non-uniformity of sound field over the surface of a manipulated object, no matter what material it is. Considering the fact that acoustic radiation force results from the non-uniformity of acoustic energy density over the surface of a manipulated object, acoustic radiation force may be generated on an object made of any material.

Also, acoustic streaming may be generated around any material because it is caused by the non-uniformity of sound field, which is an inherent property of most sound fields. The surface properties of a sample, such as surface roughness and stickiness, may influence the speed of adjacent acoustic streaming a bit, but cannot eliminate the existence of acoustic streaming.

(2) Little heat damage to manipulated samples in some methods

For a trapping based on acoustic radiation force generated by the standing wave field, the trapping is implemented near sound pressure nodes (zero sound pressure positions) for most manipulated samples, where the acoustic cavitation effect is weak. For the trapping based on acoustic streaming, the circulation of acoustic streaming carries away the heat generated by ultrasound at the manipulation location. In the trapping of single nanowires based on mobile acoustic streaming, the temperature rise at the manipulation point can be less than 0.1 ℃ [24]. More experimental work on this topic can be found in Refs.31-33.

(3) Diverse manipulation functions

The ultrasonic manipulation technology has been successfully used to trap, orientate, transfer and rotate single nanowires, to concentrate nano- and microscale or larger materials, to trap, sort and extract micro particles, to transport, generate, rotate, adsorb, split and merge droplets, and to form acoustic patterns in suspension. In the ultrasonic micro/nano trapping, existing techniques include contact and noncontact types, that is, trapped samples may be in contact or not in contact with the trapping probe. In the ultrasonic rotary driving of micro solid particles, both spin and revolution have been realized; in the ultrasonic rotary driving of nano objects, rotation of a nanowire around its central or end point has been realized.

(4) Simple and flexible device structures

The ultrasonic manipulation device usually consists of a vibration excitation unit and micro probe (or micro-fabricated manipulator). It can also be in the form of a stage. In either case, the device structure is simple and easy to fabricate and use. One may easily fabricate an acoustic manipulation device for his/her own applications without the expensive MEMS fabrication equipment.

The devices for ultrasonic micro/nano manipulations or (ultrasonic manipulators) are usually excited by piezoelectric components. Just like other piezoelectric devices such as ultrasonic motors, the structure of ultrasonic manipulators is quite flexible, which means that it can have different shapes such as needle, disk, etc. This makes the devices be easily integrated into microscope systems.

With the above features, the ultrasonic method will compete with other physical methods and serve as an alternative in practical micro/nano manipulation processes in the fields such as handling of biological samples [34-42], fabrication of microelectronic and photonic devices [43, 44], crystal growth and material fabrication [45, 46], measurement of micro/nano materials [24], transportation of micro objects [47], rotary driving of micro components [48], direction control of individual nanowires [49], infrared spectroscopy of cell suspensions [50], etc.

References

1. Ashkin, A. (1970). Acceleration and trapping of particles by radiation pressure, Phys. Rev. Lett., 24, pp. 156–159.
2. Ashkin, A., Dziedzic, J. M., Bjorkholm, J. E. and Chu, S. (1986). Observation of a single–beam gradient force optical trap for dielectric particles, Opt. Lett., 11 (5), pp. 288–290.
3. Hill, M. (1987). “He wrote the book on atom trapping,” (www.bell–labs.com/user/feature/archives/ashkin/)
4. Peterman, E. J. G., Gittes, F. and Schmidt, C. F. (2003). Laser–induced heating in optical traps, Biophys. J., 84 (2), pp. 1308–1316.
5. Petit, T., Zhang, L., Peyer, K. E., Kratochvil, B. E. and Nelson, B. J. (2012). Selective trapping and manipulation of microscale objects using mobile microvortices, Nano Lett., 12 (1), pp.156–160.
6. Tanase, M., Bauer, L. A., Hultgren, A., Silevitch, D. M., Sun, L., Reich, D. H., Searson, P. C. and Meyer, G. J. (2001). Magnetic alignment of fluorescent nanowire, Nano Lett., 1 (3), pp. 155–158.
7. Lee, S. W., Jeong, M. C., Myoung, J. M., Chae, G. S. and Chung, I. J. (2007). Magnetic alignment of ZnO nanowire for optoelectronic device applications, Appl. Phys. Lett., 90 (13), 133115.

8. Castillo, J., Dimaki, M. and Svendson, W. E. (2009). Manipulation of biological samples using micro and nano techniques, Integr. Biol., 1, pp. 30–42.
9. Fan, D. L., Zhu, F. Q., Cammarata, R. C. and Chien, C. L. (2011). Electric Tweezers, Nanotoday, 6, pp. 339–354.
10. Wang, M. C. P. and Gates, B. D. (2009). Directed assembly of nanowires, Mater. Today, 12 (5), pp. 34–43.
11. Molhave, K., Wich, T., Kortschack, A. and Boggild, P. (2006). Pick–and–place nanomanipulation using microfabricated grippers, Nature Nanotech., 17 (10), pp. 2434.
12. Carlson, K., Andersen, K. N., Eichorn, V., Petersen, D. H., Mlhave, K., Bu, I. Y. Y., Teo, K. B. K., Milne, W. I., Fatikow, S. and Bggild, P. (2007). A carbon nanofibre scanning probe assembled using an electrothermal microgripper, Nature Nanotech., 18 (34), 345501.
13. Sitti, M., Aruk, B., Shintani, K. and Hashimoto, H. (2003). Scaled teleoperation system for nano - scale interaction and manipulation, Advanced Robotic, 17 (3), 275 - 291.
14. Xie, H. and Regnier, S. (2009). Three–dimensional automated micromanipulation using a nanotip gripper with multi–feedback, IEEE/ASME. T. Mech., 19 (7), 075009.
15. Kim, S., Shafier, F., Ratchford, D. and Li, X. (2011). Controlled AFM manipulation of small nanoparticles and assembly of hybrid nanostructures, Nature Nanotech., 22 (11), pp. 115301.
16. Pihl, J., Sinclair, J., Karlsson, M. and Orwar, O. (2005). Microfluidics for cell–based assays, Mater. Today, 8 (12), pp. 46–51.
17. O'Rorke, R. D., Wood, C. D., Walti, C., Evans, S. D. and Davies, A. G. (2012). Acousto-microfluidics: Transporting microbubble and microparticle arrays in acoustic traps using surface acoustic waves, J. Appl. Phys., 111 (9), pp. 094911.
18. Coakley, W. T., Bardsley, D. W and Grundly, M. A. (1989). Cell manipulationin ultrasonic standing wave fields, J. Chem. Technol. Biotechnol., 44 (1), pp. 43 - 62.
19. Wu, J. (1991). Acoustic tweezers, J. Acoust. Soc. Am., 89 (5), pp. 2140 - 2143.
20. Hu, J. H and Santoso, A. K. (2004). A Pi–shaped Ultraso1nic Tweezers Concept for Manipulation of Small Particles, IEEE Trans. Ultrason. Ferroelectr. Freq. Control, 51 (11), pp. 1499–1507.
21. Hu, J. H., Yang, J. B and Xu, J. (2004). Ultrasonic trapping of small particles by sharp edges vibrating in a flexural mode, Appl. Phys. Lett., 85 (24), pp. 6042–6044.
22. Hu, J. H., Tay, C. L., Cai, Y. M and Du, J. L. (2005). Controlled rotation of sound–trapped small particles by an acoustic needle, Appl. Phys. Lett., 87 (9), 094104.

23. Sarvazyan, A. P.; Rudenko, O. V.; Nyborg, W. L. (2010). Biomedical applications of radiation force of ultrasound: historical roots and physical basis, Ultrasound Med. Biol., 36(9), 1379-1394.
24. Li, N., Hu, J. H., Li, H. Q., Bhuyan, S. and Zhou, Y. J. (2012). Mobile acoustic streaming based trapping and 3 - dimensional transfer of a single nanowire, Appl. Phys. Lett., 101 (9), pp. 093113.
25. Hu, J. H., Tan, C. L. and Hu, W.Y. (2007). Ultrasonic microfluidic transportation based on a twisted bundle of thin metal wires, Sensor Actuat. A - Phys., 135 (2), pp. 811–817.
26. Tan, Z. W., Teo, G. and Hu, J. H. (2008). Ultrasonic generation and rotation of a small droplet at the tip of a hypodermic needle, J. Appl. Phys., 104 (10), 104902.
27. Chung, S. K and Cho, S. K. (2008). 3–D manipulation of millimeterand - and micro–sized objects using an acoustically–excited oscillating bubble, Microfluid. Nanofluid., 6 (2), pp. 261–265.
28. Chung, S. K., Rhee, K and Cho, K. C. (2010). Bubble actuation by electrowetting - on - dielectric (EWOD) and its applications: A review, Int. J. Precis. Eng.Man., 11 (6), pp. 991 - 1006.
29. Friend, J and Yeo, L. Y. (2011). Microscale acoustofluidics: Microfluidics driven via acoustics and ultrasonics, Rev. Mod. Phys., 83 (2), pp. 647 - 704.
30. Hagiwara, M., Kawahara, T and Arai, F. (2012). Local streamline generation by mechanical oscillation in a microfluidic chip for noncontact cell manipulations, Appl. Phys. Lett., 101 (7), 074102.
31. Bazou, D., Coakley, W. T., Hayes, A. J and Jackson, S. K. (2008). Long–term viability and proliferation of alginate–encapsulated 3–D HepG2 aggregates formed in an ultrasound trap, Toxicol in Vitro., 22 (5), pp. 1321 - 1331.
32. Bazou, D., Kuznetsova, L. and Coakley, W. T. (2005). Physical enviroment of 2–D animal cell aggregates formed in a short pathlength ultrasound standing wave trap, Ultrasound Med. Biol., 31 (3), pp. 423 - 430.
33. Jinsson, H., Holm, C., Nilsson, A., Petersson, F., Johnsson, P. and Laurell, T. (2004). Particle separation using ultrasound can radically reduce embolic load to brain after cardiac surgery, Ann. Thorac. Surg., 78 (5), 1572–1577.
34. Bernassau, A. L., Courtney, C. R. P., Beeley, J., Drinkwater, B. W., and Cumming, D. R. S. (2013). Interactive manipulation of microparticles in an octagonal sonotweezer, Appl. Phys. Lett., 102, 164101.
35. Bernassau, A. L., Ong, C. K., Ma, Y., et al. (2011). Two-dimensional manipulation of micro particles by acoustic radiation pressure in a heptagon cell, IEEE Trans. Ultrason. Ferroelectr. Freq. Control, 58 (10), 2132-2138.
36. Lam, K. H., Hsu, H. S., Li, Y, et al. (2013). Ultrahigh frequency lensless ultrasonic transducers for acoustic tweezers application, Biotechnol. Bioeng., 110(3), 881-886.
37. Jeong, J. S., Lee, J. W.; Lee, Chang, Y.; et al. (2011). Particle manipulation in a microfluidic channel using acoustic trap, Biomed, Microdevices, 13(4), 779-788.

38. Benes, E., Groschl, M., Nowotny, H., Trampler, F., Keijzer, T., Bohm, H., Radel, L. G, Hawkes, J. J., Konig, R., Delouvroy, C. (2001), Ultrasonic separation of suspended particles, *IEEE Ultrason. Symp.*, pp. 649-659.
39. Orloff, N. D.; Dennis, J. R.; Cecchini, M; et al. (2011). Manipulating particle trajectories with phase-control in surface acoustic wave microfluidics, Biomicrofluidics, 5(4), 044107.
40. Guo, S. S., Zhao, L. B., Zhang, K., Lam, K. H., Lau, S. T., Zhao, X. Z., Wang, Y., Chan, H. L. W., Chen, Y., and Baigl, D. (2008). Ultrasonic particle trapping in microfluidic devices using soft lithography, Appl. Phys. Lett., 92(21), 213901.
41. Grinenko, A., Wilcox, P. D., Courtney, C. R. P. and Drinkwater, B. W. (2012). Acoustic radiation force analysis using finite difference time domain method, J. Acoust. Soc. Am., 131(5), pp.3664-3670.
42. Bruus, H., Dual, J.; Hawkes, J., et al. (2011). Forthcoming Lab on a Chip tutorial series on acoustofluidics: Acoustofluidics-exploiting ultrasonic standing wave forces and acoustic streaming in microfluidic systems for cell and particle manipulation, Lab Chip, 11(21), pp.3579-3580.
43. Vandaele, V., Lambert, P., Delchambre, A. (2005). Non-contact handling in microassembly: Acoustical levitation, Precis., Eng., 29, pp. 491-505.
44. Wang, W, Castro, L. A., Hoyos, M., et al. (2012). Autonomous Motion of Metallic Microrods Propelled by Ultrasound, ACS Nano, 6(7), pp. 6122-6132.
45. Chung, S. K., Trinh, E. H. (1998). Containerless protein crystal growth in rotating levitated drops, J. Crystal Growth, 194, pp. 384-397.
46. Geng, D. L.; Xie, W. J.; Wei, B. (2012). Containerless solidification of acoustically levitated Ni-Sn eutectic alloy, Appl. Phys A-Mat. Sci Process, 109 (1), pp. 239-244.
47. Koyama, D., Nakamura, K. (2010). Noncontact ultrasonic transportation of small objects in a circular trajectory in air by flexural vibrations of a circular disc, IEEE Trans. Ultrason. Ferroelectr. Freq. Control, 57(6), pp.1434-1442.
48. Hu, J., Zhu, X., Zhou, Y., Li, N. (2012). Principle of the Rotation of Small Particles around a Nodal Point of Strip in Flexural Vibration, Sens. Actuators A, 178, pp. 202-208.
49. Li, N., Hu, J. (2103). Sound Controlled Rotary Driving of a Single Nanowire, IEEE T NANOTECHNOL, in press.
50. Koch, C., Brandstetter, M., Lendl, B. and Radel, S. (2013). Ultrasonic Manipulation of Yeast Cells in Suspension for Absorption Spectroscopy with an Immersible Mid-Infrared Fiberoptic Probe, Ultrasound Med. Biol., 39(6), pp.1094-1101.

Chapter 2

Physics in Ultrasonic Micro/Nano Manipulations

This chapter gives major physical principles and concepts in ultrasonic micro/nano manipulations. They include piezoelectric transduction, acoustic field and energy density, acoustic radiation force, acoustic streaming, frictional driving, Chladni effect, acoustic cavitation and Bjerknes forces, and sound induced intermolecular force change. The physical principles and their proper combinations may be employed in various micro/nano manipulations. The contents in this chapter provide the fundamentals to understand operating principles of existing ultrasonic devices for micro/nano manipulations, and to design and analyze an ultrasonic manipulating device.

2.1 Piezoelectric Transduction

Ultrasonic manipulation technology makes use of physical effects of ultrasound to implement a manipulation function. The ultrasonic field for the manipulations is usually generated by a piezoelectric transducer, which converts electric energy applied to the electric input port of itself into a mechanical vibration. Piezoelectric components are the heart of the transducer. The piezoelectric transducer is usually constructed by piezoelectric components and metal parts.

2.1.1 *Piezoelectricity*

Piezoelectric effect can be divided into direct piezoelectric effect and converse piezoelectric effect. The former was discovered by French physicists Jacques Curie and Pierre Curie in 1880, and the latter was first found by Gabriel Lippmann in 1881 in his theoretical work, and experimentally confirmed by the Curies in the same year. In the direct piezoelectric effect, mechanical force applied to a piezoelectric material is converted to electrical charge on electrodes of the piezoelectric material. In the converse piezoelectric effect, electric field applied to a piezoelectric material is converted to mechanical strain [1,2]. When the temperature of a piezoelectric component is higher than certain value T_c (termed the Curie point), it loses piezoelectricity. The Curie point of most commercialized piezoelectric components is around 300℃. To ensure a reliable operation, the temperature of a piezoelectric component during operation should be limited below half of the Curie point.

Piezoelectric materials may be natural and synthetic. Natural piezoelectric materials include some crystals such as quartz, rochelle salt, table sugar and berlinite, and some biological materials such as human bones, proteins, etc. Synthetic piezoelectric materials include the ceramics such as lead zirconate titanate (commonly known as PZT), zinc oxide (ZnO), barium titanate ($BaTiO_3$), sodium potassium niobate ($(K, Na)NbO_3$), bismuth ferrite ($BiFeO_3$), polymers such as polyvinylidene fluoride (PVDF), crystals such as gallium orthophosphate ($GaPO_4$), etc. Ultrasonic manipulation devices mainly use lead zirconate titanate (PZT) materials due to the considerations in vibration strength, fabrication process and device cost.

Piezoelectric constitutive equations define the relationships among the stress matrix **T**, strain matrix **S**, electric displacement or charge density matrix **D**, and electric field matrix **E** of a piezoelectric component. The piezoelectric constitutive equation in strain-charge form is

$$\mathbf{S} = \mathbf{s_E} \cdot \mathbf{T} + \mathbf{d^t} \cdot \mathbf{E} \quad (2.1.1)$$

$$\mathbf{D} = \mathbf{d} \cdot \mathbf{T} + \boldsymbol{\varepsilon}_\mathbf{T} \cdot \mathbf{E} \quad (2.1.2)$$

where $\mathbf{s_E}$ is the compliance coefficient matrix measured at zero electric field, **d** is the piezoelectric coefficient matrix for the strain-charge form, $\boldsymbol{\varepsilon_T}$ is the permittivity coefficient matrix measured at zero stress, and superscript t stands for a transposed matrix. For a poled PZT component, Eqs. (2.1.1) and (2.1.2) can be extended into the following forms:

$$\begin{bmatrix} S_1 \\ S_2 \\ S_3 \\ S_4 \\ S_5 \\ S_6 \end{bmatrix} = \begin{bmatrix} s_{11}^E & s_{12}^E & s_{13}^E & 0 & 0 & 0 \\ s_{21}^E & s_{22}^E & s_{23}^E & 0 & 0 & 0 \\ s_{31}^E & s_{32}^E & s_{33}^E & 0 & 0 & 0 \\ 0 & 0 & 0 & s_{44}^E & 0 & 0 \\ 0 & 0 & 0 & 0 & s_{55}^E & 0 \\ 0 & 0 & 0 & 0 & 0 & s_{66}^E = 2\left(s_{11}^E - s_{12}^E\right) \end{bmatrix} \begin{bmatrix} T_1 \\ T_2 \\ T_3 \\ T_4 \\ T_5 \\ T_6 \end{bmatrix} + \begin{bmatrix} 0 & 0 & d_{31} \\ 0 & 0 & d_{32} \\ 0 & 0 & d_{33} \\ 0 & d_{24} & 0 \\ d_{15} & 0 & 0 \\ 0 & 0 & 0 \end{bmatrix} \begin{bmatrix} E_1 \\ E_2 \\ E_3 \end{bmatrix} \tag{2.1.3}$$

$$\begin{bmatrix} D_1 \\ D_2 \\ D_3 \end{bmatrix} = \begin{bmatrix} 0 & 0 & 0 & 0 & d_{15} & 0 \\ 0 & 0 & 0 & d_{24} & 0 & 0 \\ d_{31} & d_{32} & d_{33} & 0 & 0 & 0 \end{bmatrix} \begin{bmatrix} T_1 \\ T_2 \\ T_3 \\ T_4 \\ T_5 \\ T_6 \end{bmatrix} + \begin{bmatrix} \varepsilon_{11} & 0 & 0 \\ 0 & \varepsilon_{22} & 0 \\ 0 & 0 & \varepsilon_{33} \end{bmatrix} \begin{bmatrix} E_1 \\ E_2 \\ E_3 \end{bmatrix} \tag{2.1.4}$$

in which 3 represents the poling direction, and 1 and 2 the two directions perpendicular to the poling direction and normal each other. The subscripts in the stress and strain matrices are given in the following conventions: $T_1 = T_{11}$; $T_2 = T_{22}$; $T_3 = T_{33}$; $T_4 = T_{23}$; $T_5 = T_{13}$; $T_6 = T_{12}$; $S_1 = S_{11}$; $S_2 = S_{22}$; $S_3 = S_{33}$; $S_4 = S_{23}$; $S_5 = S_{13}$; $S_6 = S_{12}$. The strains S_{ij} (i, j = 1, 2, or 3) are related to the displacements u_1, u_2 and u_3 in the directions 1, 2 and 3 by the following equations:

$$S_{ij} = \frac{1}{2}\left(\frac{\partial u_i}{\partial x_j} + \frac{\partial u_j}{\partial x_i}\right) \tag{2.1.5}$$

where i, j = 1, 2 or 3, and x_i and x_j are the coordinates in the $x_1x_2x_3$ coordinate system. The positive sign of strain S_k (k = 1, 2, or 3) of a piezoelectric element means that the element extends along the *k* direction, and the negative sign means it contracts along the *k* direction.

S, **T**, **D**, and **E** can be rearranged to give other forms for a piezoelectric constitutive equation. The stress-charge and strain-voltage forms are also often used in engineering calculation and analyses. They are as follows:

Stress-charge form:

$$\mathbf{T} = \mathbf{c_E} \cdot \mathbf{S} - \mathbf{e^t} \cdot \mathbf{E} \tag{2.1.6}$$

$$\mathbf{D} = \mathbf{e} \cdot \mathbf{S} + \boldsymbol{\varepsilon}_\mathbf{s} \cdot \mathbf{E} \tag{2.1.7}$$

Strain-voltage form:

$$\mathbf{S} = \mathbf{s_D} \cdot \mathbf{T} + \mathbf{g^t} \cdot \mathbf{D} \tag{2.1.8}$$

$$\mathbf{E} = -\mathbf{g} \cdot \mathbf{T} + \boldsymbol{\varepsilon}_\mathbf{T}^{-1} \cdot \mathbf{D} \tag{2.1.9}$$

The material property parameter matrices **d**, **e**, **g**, $\mathbf{s_E}$, $\mathbf{s_D}$, $\mathbf{c_E}$, $\boldsymbol{\varepsilon}_\mathbf{s}$, and $\boldsymbol{\varepsilon}_\mathbf{T}$ can be transformed by the following relationships:

$$\mathbf{c_E} = \mathbf{s_E^{-1}} \tag{2.1.10}$$

$$\mathbf{c_D} = \mathbf{s_D^{-1}} \tag{2.1.11}$$

$$\mathbf{c_D} = \mathbf{c_E} + \mathbf{e^t} \cdot \boldsymbol{\varepsilon}_\mathbf{S}^{-1} \cdot \mathbf{e} \tag{2.1.12}$$

$$\mathbf{s_D} = \mathbf{s_E} - \mathbf{d^t} \cdot \boldsymbol{\varepsilon}_\mathbf{T}^{-1} \cdot \mathbf{d} \tag{2.1.13}$$

$$\boldsymbol{\varepsilon}_\mathbf{s} = \boldsymbol{\varepsilon}_\mathbf{T} - \mathbf{d} \cdot \mathbf{s_E^{-1}} \mathbf{d^t} \tag{2.1.14}$$

$$\boldsymbol{\varepsilon}_\mathbf{S}^{-1} = \boldsymbol{\varepsilon}_\mathbf{T}^{-1} + \mathbf{g} \cdot \mathbf{s_D^{-1}} \cdot \mathbf{g^t} \tag{2.1.15}$$

$$\mathbf{e} = \mathbf{d} \cdot \mathbf{s_E^{-1}} \tag{2.1.16}$$

$$\mathbf{g} = \boldsymbol{\varepsilon}_\mathbf{T}^{-1} \cdot \mathbf{d} \tag{2.1.17}$$

More details about the piezoelectric constitutive equations and their applications can be found in Refs 1-4.

2.1.2 *Piezoelectric components and their vibration modes*

Using piezoelectric components in a transducer can make the vibration excitation structure more efficient and compact. Figure 2.1.1 shows the images of some piezoelectric components, which have been used or have potential applications in ultrasonic manipulation devices. The piezoelectric components may have various shapes such as disk, ring, rectangle, arc, etc., which makes the design of ultrasonic manipulation devices very flexible. To make the devices compact in structure, piezoelectric components with low profile such as ring, disk and plate type components, are usually used, and the piezoelectric components may be bonded onto the surface of a metal part in a vibration excitation structure by adhesive material.

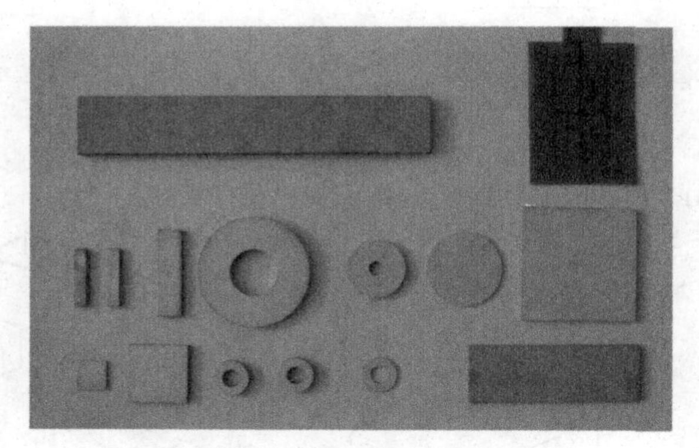

Fig. 2.1.1. Piezoelectric components.

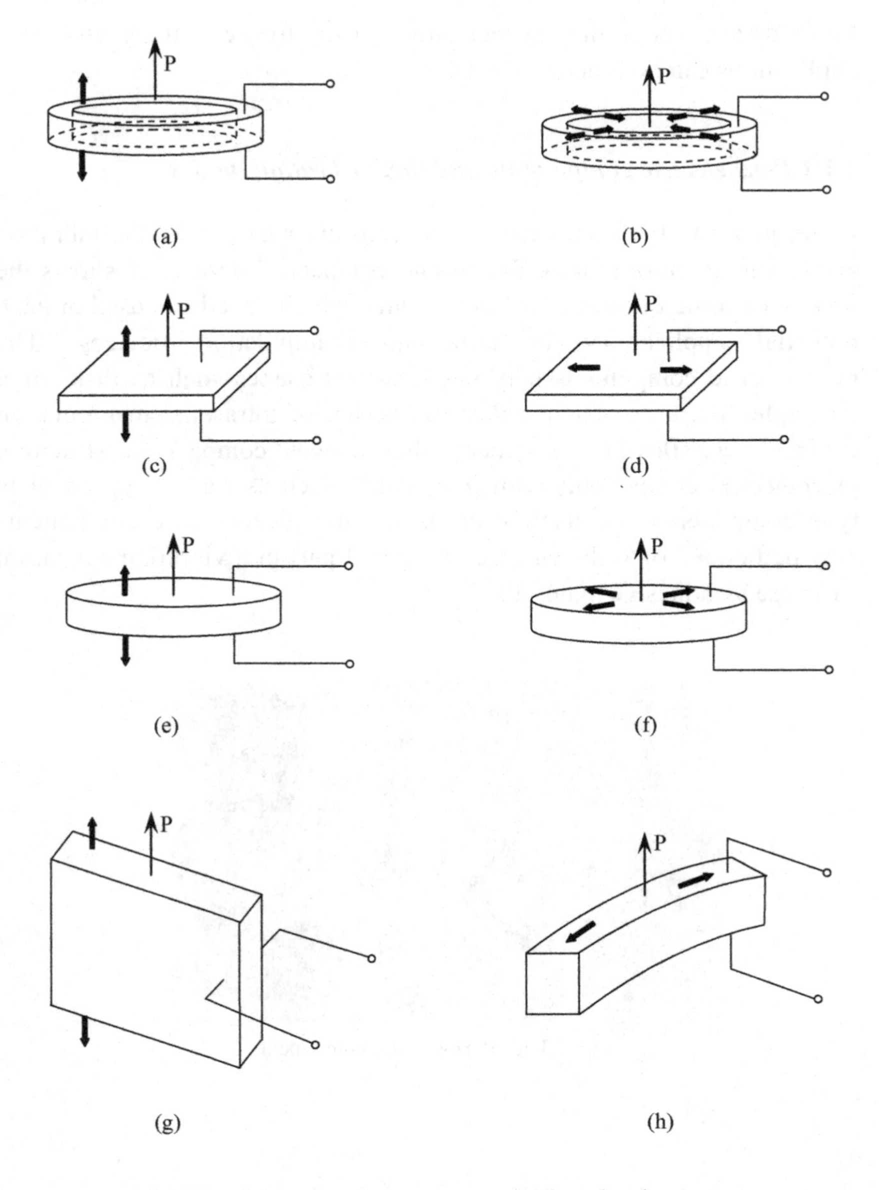

Fig. 2.1.2. Vibration modes of some commonly piezoelectric components.

Some commonly used vibration modes of piezoelectric components are shown in Fig. 2.1.2. In the figures, the top and bottom surfaces of each component are covered with whole electrodes (usually made of silver), and a driving voltage or electric field is applied to the two opposite electrodes of each component. The polarization direction of each component is represented by the *P* axis. The thicker arrows represent the vibration direction.

In Figs. 2.1.2(a) and (b), there is a piezoelectric ring poled in the thickness direction, and the driving electric field is in the thickness direction. In Fig. 2.1.2(a), the ring vibrates in the thickness vibration mode, that is, the ring extends and contracts in the thickness direction. Electromechanical transduction capability of the piezoelectric components operating in this mode may be evaluated by electromechanical coupling factor k_t of the piezoelectric material, sometimes provided by its producer. The thickness mode of a piezoelectric ring is widely used in sandwich type transducers which are well used in vibration excitation of the ultrasonic probes and radiation faces for manipulating micro/nano particles. In Fig. 2.1.2(b), the same piezoelectric ring operates in the wall thickness mode. Electromechanical transduction capability of the piezoelectric components operating in this mode may be evaluated by electromechanical coupling factor k_{31} of the piezoelectric material. This mode can be used in the vibration excitation of an ultrasonic stage.

In Figs. 2.1.2(c) and (d), there is a rectangular piezoelectric plate poled in the thickness direction, and the driving electric field is in the thickness direction. It operates in the thickness vibration mode in Fig. 2.1.2(c), and in the extensional mode in Fig. 2.1.2(d). Electromechanical transduction capability of the former mode can be evaluated by electromechanical coupling factor k_t of the material, and that of the latter by the electromechanical coupling factor k_{31}. The thickness mode of rectangular piezoelectric plates may be used in longitudinal and bending type Langevin transducers, and the extensional mode is very useful in the flexural vibration excitation mechanism such as the bimorph piezoelectric vibrator.

In Figs. 2.1.2(e) and (f), there is a piezoelectric disk poled in the thickness direction, and the driving electric field is in the thickness direction. The disk vibrates in the thickness mode. Electromechanical transduction capability of the piezoelectric components operating in these two modes can be evaluated by electromechanical coupling factors of the material k_t and k_r (or k_p), respectively. The thickness mode of a piezoelectric disk may be used to generate an ultrasonic field if the piezoelectric disk is bonded onto a radiation face, and the radial mode may also be used to excite the flexural vibration of an ultrasonic stage.

In Fig. 2.1.2(g), there is a piezoelectric plate poled in the width direction, and the applied electric field is in the thickness direction. The piezoelectric plate vibrates in the thickness shear mode. Electromechanical transduction capability of the piezoelectric components operating in this mode may be evaluated by the electromechanical coupling factor k_{15}. In Fig. 2.1.2(g), there is an arc-shaped piezoelectric component poled in the thickness direction, which may be cut from a piezoelectric whole ring, and the driving electric field is in the thickness direction. The piezoelectric arc operates in the extensional vibration mode. Electromechanical transduction capability of the piezoelectric components operating in this mode may be evaluated by the electromechanical coupling factor k_{31}. The extensional mode of a piezoelectric arc may be used in an ultrasonic stage to generate a traveling wave around nodal points.

In addition to the above listed modes, piezoelectric components have other vibration modes. In the length vibration mode, a piezoelectric bar or cylinder polarized in its length direction vibrates in the polarized direction with the driving electric field parallel to the polarized direction. Electromechanical transduction capability of the piezoelectric components operating in this mode may be evaluated by k_{33} of the material. More details of the vibration modes of piezoelectric components can be found in Refs.5-7.

The piezoelectric ring, disk and plate to generate the thickness mode shown in Figs. 2.1.2 (a), (c) and (e) can be very thin in order to excite a stand wave acoustic field in a micro chamber for micro particle agglomeration. For the commercialized piezoelectric components, k_{31} is usually the smallest among these electromechanical coupling factors. But in many applications, using the k_{31} mode can make devices possess a low profile and simple structure. In addition, many cases of ultrasonic micro/nano manipulations do not need a strong vibration. Thus it is possible to use the lead free piezoelectric components in the devices.

2.1.3 *Langevin transducers*

The Langevin transducer is an important ultrasonic transducer, named after a famous French physicist Paul Langevin (1872-1946). It usually consists of the front mass, piezoelectric rings, back mass and bolting structure. And the piezoelectric rings are sandwiched between the front and back masses by the bolting structure. Commercialized Langevin transducers can also be used in ultrasonic manipulation devices. Fig. 2.1.3 shows the images of some Langevin transducers. The resonance frequency of the transducers from the right hand side to the left hand side is 20.5 kHz, 25.2 kHz, 47.1 kHz, 66.8 kHz, 123.4 kHz and 198.1 kHz, respectively. As a whole, the transducer size increases as its resonance frequency decreases if the same materials are used for the metal parts. For a transducer with fixed resonance frequency, using metal material with larger density can decrease the transducer size. A probe for ultrasonic micro/nano manipulations may be bonded or clamped onto the radiation surface of a Langevin transducer to be mechanically excited. Main merits of using a Langevin transducer in ultrasonic manipulating devices include large vibration and high reliability. However, using a Langevin transducer may increase the volume and mass of the manipulating system.

Fig. 2.1.3. Langevin transducers operating in the longitudinal mode.

2.2 Acoustic Field and Energy Density

2.2.1 *Basic concepts and wave equations*

Most of ultrasonic manipulations rely on ultrasonic field generated by the devices. For an ultrasonic field outside the acoustic boundary layer, the steady state solutions of its first-order pressure p_1 (= sound pressure p) and first-order velocity v_1 (= vibration velocity v) can be solved or numerically calculated from the wave equation and its boundary conditions [8]. The wave equation is

$$\frac{\partial^2 \varphi}{\partial t^2} = c_0^2 \nabla^2 \varphi \tag{2.2.1}$$

where φ is the velocity potential, c_0 is the sound speed in the acoustic medium, and ∇^2 is the Laplacian operator ($\equiv$ div grad). In the Cartesian coordinate system, ∇^2 is

$$\nabla^2 = \frac{\partial^2}{\partial \mathrm{x}^2} + \frac{\partial^2}{\partial \mathrm{y}^2} + \frac{\partial^2}{\partial \mathrm{z}^2} \tag{2.2.2}$$

The relationship between the vibration velocity v and velocity potential φ is

$$\boldsymbol{v} = -\mathrm{grad}\,\varphi \tag{2.2.3}$$

The relationship between the sound pressure p and velocity potential φ is

$$p = \rho_0 \frac{\partial \varphi}{\partial t} \tag{2.2.4}$$

where ρ_0 is the density of acoustic medium when there is no sound. At a rigid boundary,

$$v_n = 0 \ \text{ or } \ \frac{\partial p}{\partial n} = 0 \tag{2.2.5}$$

which means the out-of- plane vibration velocity at the boundary is zero. At a soft sound boundary such as the liquid-gas interface for an acoustic field in liquid,

$$p = 0 \tag{2.2.6}$$

At a radiation face with an out-of -plane vibration velocity v_s (t, x, y, z),

$$v_n = v_s(t, x, y, z) \tag{2.2.7}$$

The wave equation can also be in the form:

$$\frac{\partial^2 p}{\partial t^2} = c_0^2 \nabla^2 p \tag{2.2.8}$$

With the solutions of vibration velocity $\boldsymbol{v}$ and sound pressure p, the kinetic energy density K and potential energy density P can be calculated by

$$K = \frac{1}{2}\rho_0 v^2 \tag{2.2.9}$$

$$P = \frac{1}{2}\frac{p^2}{\rho_0 c_0^2} \tag{2.2.10}$$

where v is the absolute value of the vibration velocity. The total energy density is the sum of K and P. The SI unit of K and U is J/m^3. The instantaneous intensity of sound or sound power flux per unit area $\boldsymbol{I}$ can be calculated by

$$\boldsymbol{I} = p\boldsymbol{v} \tag{2.2.11}$$

2.2.2 *Standing and traveling waves*

The standing and traveling waves are two special types of acoustic fields, and have wide and important applications in ultrasonic micro/nano manipulations. A monochromatic standing wave can be expressed by

$$p = p_m \sin(\omega t + \alpha)\sin(k_1 x + \beta_1)\sin(k_2 y + \beta_2)\sin(k_3 z + \beta_3) \tag{2.2.12}$$

where p_m is the amplitude, ω the angular frequency, and k_1, k_2 and k_3 the wave numbers. In practice, p_m usually changes with the field position. The basic feature of a monochromatic standing wave is that its phase angle does not change with space.

A monochromatic traveling wave along the x axis can be expressed by

$$p = p_m \sin(\omega t \pm kx + \alpha) \tag{2.2.13}$$

where k is the wave number, the positive sign before kx means the wave travels to the $-x$ direction, and the negative sign means the wave travels to the $+x$ direction. In practical devices, the amplitude p_m is usually a function of space. The basic feature of a monochromatic traveling wave is that its phase changes with the position of field point linearly.

2.2.3 *Viscous boundary layer of acoustic field*

The rotational or transverse solution becomes important for the acoustic field near a boundary [8]. The thickness of viscous boundary layer of an acoustic field is

$$\delta = \sqrt{\frac{2\eta}{\omega \rho_0}} \tag{2.2.14}$$

where η and ρ_0 are the shear viscosity and density of the acoustic medium, respectively, and ω the angular frequency of the acoustic field. The viscous boundary layer of an acoustic field usually is very thin; e.g. it is 3.6 μm and 14.3 μm thick at 25 kHz and room temperature in water and air, respectively. In this viscous boundary layer, vibration velocity of the acoustic medium is

$$\boldsymbol{v} = \boldsymbol{v}_l + \boldsymbol{v}_t \tag{2.2.15}$$

where $\boldsymbol{v}_l$ is the longitudinal component of the vibration velocity and $\boldsymbol{v}_t$ the rotational or transverse component.

The longitudinal part can be solved from the following equations:

$$\rho_0 \frac{\partial \boldsymbol{v}_l}{\partial t} = -\text{grad}p + \left(\eta + \frac{4}{3}\mu\right) \nabla^2 \boldsymbol{v}_l \tag{2.2.16}$$

$$\text{curl}\, \boldsymbol{v}_l = 0 \tag{2.2.17}$$

where μ is the volume viscosity of the acoustic medium. For the acoustic medium which is not very sticky, Eqs. (2.2.16) and (2.2.17) can be transformed into Eq. (2.2.1) or (2.2.8).

The transverse part can be solved from the following equation:

$$\rho_0 \frac{\partial \boldsymbol{v}_t}{\partial t} = -\mu \text{curl}\,(\text{curl}\,\boldsymbol{v}_t) \qquad (2.2.18)$$

$$\text{div}\,\boldsymbol{v}_t = 0 \qquad (2.2.19)$$

The transverse part obeys a diffusion equation. It attenuates rapidly in the boundary layer and thus is important only within the boundary.

The acoustic streaming eddy in the viscous boundary layer of an acoustic field, also known as near-boundary acoustic streaming, has been well researched [59-61]. Due to a large spatial gradient of the near-boundary acoustic streaming, a substantial shear force on a micro/nano object in the boundary can be generated by the streaming. The force has been used to rotary or linearly drive a rotor or slider [61-66], and to stabilize an acoustically levitated object [67].

2.3 Acoustic Radiation Force

2.3.1 *Generality*

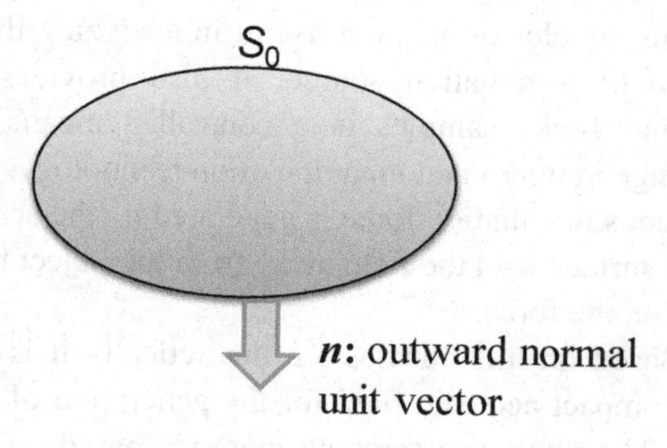

Fig. 2.3.1. An object experiencing acoustic radiation force.

Acoustic radiation force is a force on an object in acoustic field, which is caused by the kinetic and potential energy densities of the acoustic field on the object surface, and momentum transfer of the acoustic field to the object. For an object in arbitrary acoustic field, as shown in Fig. 2.3.1, acoustic radiation force on it is [9, 10]

$$\boldsymbol{F} = \left\langle \iint_{S_0} (K - P)\boldsymbol{n} dS \right\rangle - \left\langle \iint_{S_0} \rho_0 \boldsymbol{v}\, v_n dS \right\rangle \tag{2.3.1}$$

where both integrals are over the surface S_0 of the object without sound, $\boldsymbol{n}$ is the outward normal unit vector of S_0, ρ_0 is the density of acoustic medium when there is no sound, $\boldsymbol{v}$ is the vibration velocity of acoustic medium on the surface S_0, v_n is the component of $\boldsymbol{v}$ in the $\boldsymbol{n}$ direction and $\langle\ \rangle$ represents the average over one or several time periods. From Eq. 2.3.1, there are the following conclusions:

(1) The acoustic radiation force results from the energy densities on the object surface and momentum transfer rate per unit area from the acoustic field to the object.

If the object is rigid, there is no momentum transfer from the acoustic field to the object. In this case, sound pressure and the tangential vibration velocity can exist on the object surface. If the object is acoustically soft, the velocity on its surface is not zero, and there is the momentum transfer.

(2) The kinetic energy density causes a force pulling the object, while the potential energy density causes a force pushing the object.

This conclusion is quite useful in analyzing the trapping by the surface of a radiation source. It also provides a guideline for releasing sticky samples in a controlled means, which is still a challenge in micro/nano manipulation technology.

(3) The acoustic radiation force is generated by the acoustic field on the object surface, and the field away from the object has no direct effect on the force.

Based on this conclusion, theoretically it is possible to use a very compact acoustic field for the generation of acoustic radiation force. The ultrasonic trapping methods based on a radiation point, line and face are the utilization of this conclusion, and confirm this conclusion.

(4) The shape of a manipulated object and its relative orientation to the acoustic field affect the acoustic radiation force.

In Ref. 10, ultrasonic field generated by two parallel radiation lines is used to investigate the effects of the shape and orientation of a trapped particle on the acoustic radiation force on it. Both experiment and computation in it confirmed this conclusion. For the particles with the same density and volume but different shapes and manipulated by the device with the same vibration, the acoustic radiation force is greatest to least for the particle shapes rectangular cuboid, cylinder, cone, cube, sphere, and hollow cylinder. A rectangular cuboid particle has six possible orientations during trapping. And the acoustic radiation force is different for these possible orientations for the same vibration of the device. The longer the action line, the larger the acoustic radiation force.

When Eq. (2.3.1) is used to calculate the acoustic radiation force on an object, the finite element method may be used to calculate the acoustic field and energy densities first [10]. Gor'kov theory provides a convenient method for estimation and analysis of the acoustic radiation force on a spherical object satisfying the condition $kR \ll 1$ (k is the wave number of acoustic field, and R the radius of manipulated spherical objects [11]. Based on the theory, acoustic radiation force acting on a particle in an acoustic field is

$$\vec{F} = -\nabla U \tag{2.3.2}$$

where U is the time-averaged force potential of the acoustic field. When the wave number k and the radius of particle R satisfy $kR \le 1$, the force potential U is

$$U = V[-D < K > +(1-\gamma) < P >] \tag{2.3.3}$$

where V is the volume of the particle sphere, $\langle K \rangle$ and $\langle P \rangle$ are the time-averaged kinetic and potential energy densities of the acoustic field, respectively, D is a parameter determined by the densities of the particle sphere and fluid, and γ is the compressibility ratio between the particle sphere and fluid. D and γ can be calculated by

$$D = \frac{3(\rho_s - \rho_0)}{2\rho_s + \rho_0} \tag{2.3.4}$$

$$\gamma = \frac{\rho_0 c_0^2}{\rho_s c_s^2} \tag{2.3.5}$$

where ρ_s and ρ_0 are the densities of the particle sphere and fluid, respectively, and c_s and c_0 are the sound speed in the particle sphere and fluid, respectively.

In practical applications, acoustic radiation force may be generated by a traveling wave sound field with a large spatial gradient [12-16], two traveling waves with opposite phase [17], a standing wave sound field [18-24], focused and Bessel sound beams [25, 26] and acoustic bubbles [27-29]. An acoustic radiation force can also be generated by a vibrating bubble and standing surface acoustic wave (SSAWs) [29-31].

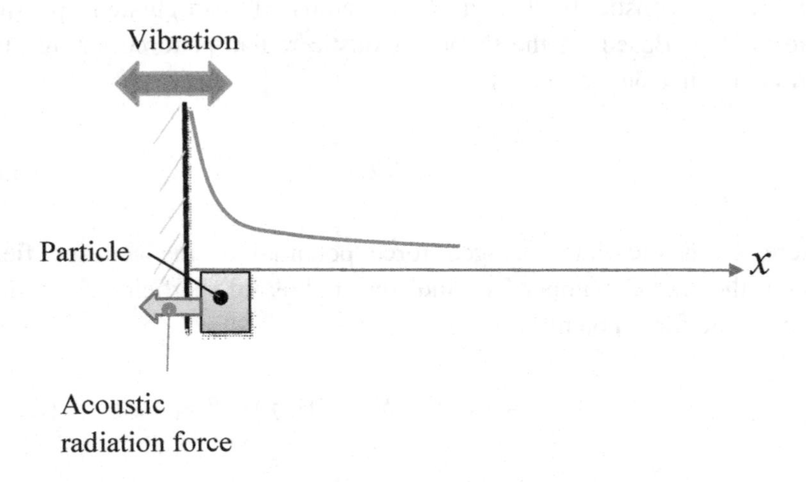

Fig. 2.3.2. Acoustic radiation force on a micro/nanoscale object in the vicinity of a radiation face.

2.3.2 *Acoustic radiation force resulting from a traveling wave with large spatial gradient*

A traveling wave with a large spatial gradient may exist in the vicinity of a radiation point, line and face, or at the leaky boundary of a standing wave. Next the acoustic radiation force on a micro/nanoscale object in the vicinity of a radiation face, as shown in Fig. 2.3.2, will be demonstrated and discussed, based on the following assumptions: there is an acoustic radiation face with uniformly distributed vibration velocity at $x = 0$; the acoustic field along the x axis is one dimensional; there is no reflection boundary along the propagation direction or the reflection boundary is far away from the radiation face.

Numerical calculation by the finite element method has shown that the sound pressure along the x-axis can be approximately expressed by [12]

$$p = p_m e^{-\alpha x} \cos(kx + \varphi_0 + \omega t) \tag{2.3.6}$$

The relationship between the acoustic pressure and vibration velocity in the sound field is

$$\rho_0 \frac{\partial v}{\partial t} = -\frac{\partial p}{\partial x} \tag{2.3.7}$$

where ρ_0 is the density of the fluid. Therefore the x-directional vibration velocity of the fluid in the sound field is

$$v = -\frac{P_m^2 \sqrt{\alpha^2 + k^2}}{\rho_f \omega} e^{-\alpha x} \cos(kx + \omega t + \theta) \tag{2.3.8}$$

where $\theta = \tan^{-1} \alpha / k$. From Eqs. (2.2.9), (2.2.10), and (2.3.2)-(2.3.5), the acoustic radiation force acting on a particle in the sound field can be derived. It is

$$\vec{F} = -\vec{i}\,\frac{V\alpha\, p_m^2 e^{-2\alpha x}}{2\rho_0}\left[\frac{D(\alpha^2 + k^2)}{\omega^2} + \frac{1-\gamma}{c_0^2}\right] \tag{2.3.9}$$

In air, $D \approx 1.5$ and $\gamma \approx 0$ because $\rho_s >> \rho_0$ for a solid particle. Thus the acoustic radiation force in air can be simplified into

$$\vec{F} = -\vec{i}\,\frac{V\alpha\, p_m^2 e^{-2\alpha x}}{4\rho_0 c_0^2} \tag{2.3.10}$$

From the above results, the following conclusions can be obtained.

- The acoustic radiation force is proportional to the square of sound pressure on the radiation face.
- The acoustic radiation force is zero if sound field adjacent to the radiation face has no spatial gradient.
- In air, the acoustic radiation force on a micro/nanocale object in the vicinity of a radiation force points to the radiation face. Thus the particle can be sucked to the radiation face.
- The acoustic radiation force increases with the increase of volume of the particle.

2.3.3 *Acoustic radiation force resulting from a standing wave*

The acoustic radiation force on micro/nanoscale objects in a standing wave acoustic field can be solved by a similar derivation. As shown in Fig. 2.3.3, in a one dimensional standing wave sound field, there are nodal and anti-nodal planes of sound pressure. The sound pressure is zero and maximum at the nodal and anti-nodal planes, respectively.

Based on the derivation, if the sound pressure is

$$p = p_m \sin(kx)\sin(\omega t + \varphi_0) \tag{2.3.11}$$

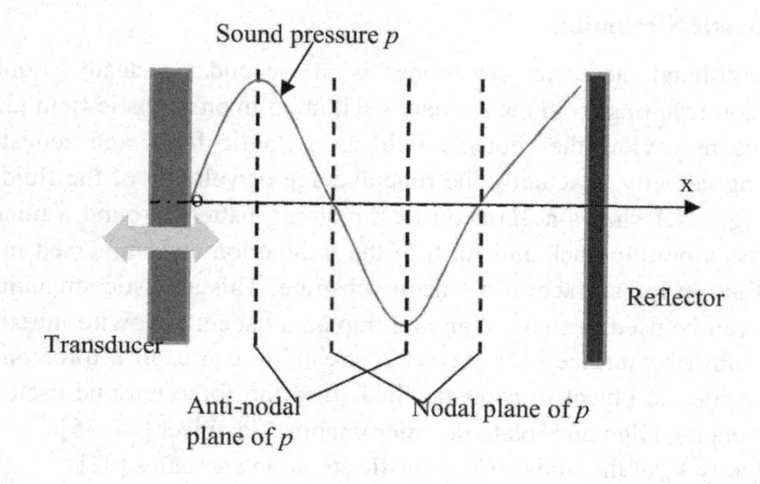

Fig. 2.3.3. A standing wave acoustic field.

then the acoustic radiation force on a particle with volume V, parallel to the x axis, is

$$F_x = \frac{-V(D+1-\gamma)kp_m^2}{4\rho_0 c_0^2}\sin 2kx \tag{2.3.12}$$

Using Eqs. (2.3.11) and (2.3.12), the following conclusions can be obtained. If the acoustic field and manipulated objects satisfies

$$D + 1 - \gamma > 0, \tag{2.3.13}$$

the acoustic radiation force drives the objects to the radiation face. Most solid objects satisfy this condition. For air bubbles in water, $D + 1 - \gamma$ is negative, and the acoustic radiation force pushes the air bubble near the radiation face away. Based on this analysis, it is known that the concentration of micro bubbles on a radiation face, which is sometimes observed in ultrasonic experiments, should not be caused by the acoustic radiation force generated by a standing wave.

2.4 Acoustic Streaming

The traditional acoustic streaming is a secondary steady liquid circulation resulting from the primary oscillation in an acoustic field [32, 33]. One may view the acoustic field as a fluidic field, and acoustic streaming velocity is actually the time average of velocity of the fluidic field. Fig. 2.4.1 shows a 3D acoustic streaming pattern around a micro fiberglass vibrating back and forth in the x direction and immersed in a water film on the surface of a silicon substrate. This acoustic streaming pattern can be used to suck, align and trap an adjacent nanowire initially on the substrate surface [34]. Acoustic streaming can exert a force on a micro/nanoscale object in acoustic field. And the force may be used to trap, orientate, align and rotate the micro/nanoscale object [34, 35].

Velocity $\boldsymbol{V_a}$ of the traditional acoustic streaming satisfies [32]

$$\mu\nabla\times\nabla\times\boldsymbol{V_a}-\left(\frac{4}{3}\mu+\mu'\right)\text{grad div}\boldsymbol{V_a}=-\nabla p_2+\boldsymbol{F} \quad (2.4.1)$$

where μ and μ' are the shear and bulk viscosity of the acoustic medium, respectively, ∇ is the Hamiltonian operator $(=\boldsymbol{i}\frac{\partial}{\partial x}+\boldsymbol{j}\frac{\partial}{\partial y}+\boldsymbol{k}\frac{\partial}{\partial z})$, p_2 is the time-independent second order sound pressure, and $\boldsymbol{F}$ is the driving force of the acoustic streaming and can be calculated from the first-order vibration velocity by

$$\boldsymbol{F}\equiv-\rho_0\langle(\boldsymbol{v}\cdot\nabla)\boldsymbol{v}+\boldsymbol{v}(\nabla\cdot\boldsymbol{v})\rangle \quad (2.4.2)$$

where $<>$ represents the time average over one period. Because div$\boldsymbol{V_a}$ is negligible, Eq. 2.4.1 can be simplified into

$$\boldsymbol{F}-\nabla p_2+\mu\Delta\mathbf{V_a}=0 \quad (2.4.3)$$

where $\Delta=\nabla\bullet\nabla=\nabla^2$. The above equations are discussed as follows:

- From Eq. 2.4.2, it is known that the generation of acoustic streaming is related to the spatial gradient of the momentum transfer rate or Reynolds stress [36] in acoustic field.
- Considering the second-order pressure p_2 is proportional to the square of the sound pressure p (the first-order pressure) which is proportional

to the vibration velocity *v* (the first-order velocity), acoustic streaming velocity V_a must be proportional to the square of vibration velocity of acoustic field.

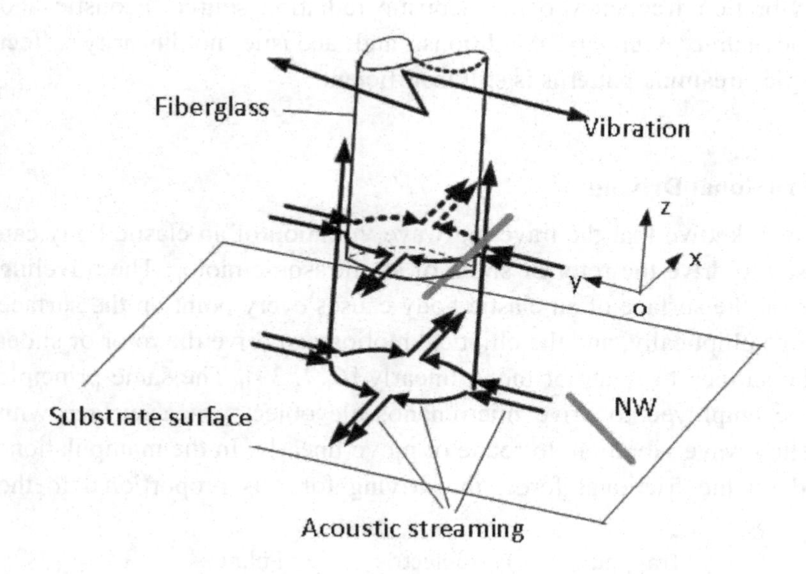

Fig. 2.4.1. A 3D acoustic streaming pattern.

It is known that the collapse of a vibrating bubble generates high-speed jets [27,28], and elliptical stir with a micro rod ultrasonically driven also causes a flow. They are also called acoustic streaming. Because the principles of these flows are different from that of the traditional one, their analysis methods should be different from the one for the traditional acoustic streaming. In addition, experimental results show that the acoustic streaming in a practical device can only occur when its acoustic field is strong enough. This phenomenon is caused by the adhesion between acoustic medium and boundaries, and among the molecules of acoustic medium.

Experimental observation and theoretical calculation have shown lots of acoustic streaming patterns [34-38]. This diversity in acoustic streaming patterns may provide possible effective principles to meet the various manipulation requirements in the fabrication of microelectronic and photonic devices, package and assembling of micro/nano devices, measurement of micro/nanoscale materials, handling of biological samples, etc. However at present, understanding of how the shape, size and vibration frequency of an acoustic radiation source, acoustic and hydrodynamic boundary conditions, and acoustic nonlinearity affect acoustic streaming patterns is still insufficient.

2.5 Frictional Driving

It is well known that the traveling wave vibration of an elastic body can be used to drive the rotor or slider of an ultrasonic motor. The traveling wave on the surface of an elastic body causes every point on the surface to move elliptically, and the elliptical motion can drive the rotor or slider on the surface to rotate or move linearly [6, 7, 39]. The same principle can be employed to drive micro/nanoscale objects on a surface with traveling wave vibration, to rotate or move linearly. In the manipulations based on the frictional force, the driving force is proportional to the

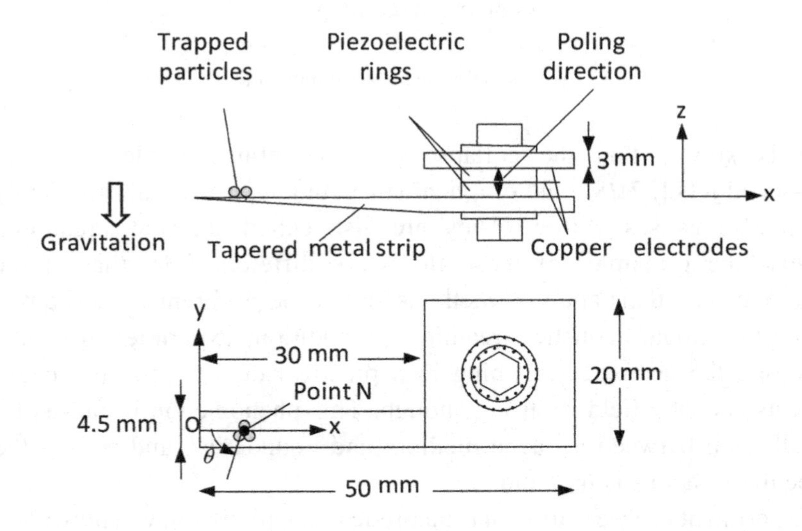

Fig. 2.5.1. A tapered metal strip excited by a Langevin transducer.

friction coefficient between the driven object and driving surface, and increases as the out-of-plane vibration of driving surface increases; the driving speed increases as the in-plane vibration of the driving surface increases.

In the ultrasonic manipulation device, there are four methods to generate the traveling wave in an elastic body. The first one is to utilize the asymmetric or non-uniform vibration excitation of an elastic plate to generate traveling waves around nodal points of the elastic plate. The second one is to dispose a vibration excitation unit at one end of an elastic plate and an absorbing unit at another end to generate a traveling wave in the plate. The third method is to use the superposition of two standing waves with spatial and temporal phase differences of $\pi/2$ and the same vibration direction, to generate a traveling wave. The fourth method is to excite two spatially orthogonal vibrations on a driving

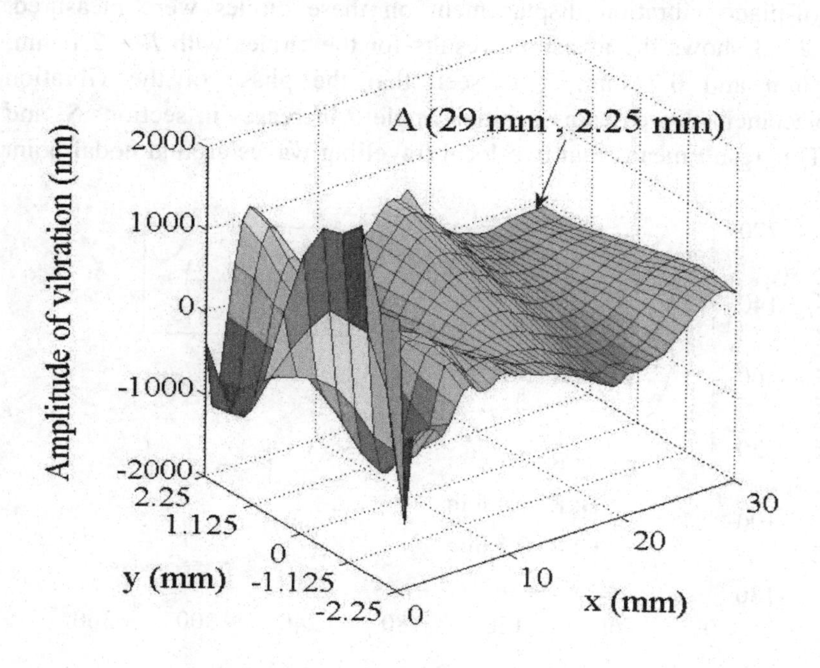

Fig. 2.5.2. Measured vibration distribution of the tapered metal plate.

surface with a temporal phase difference of $\pi/2$. The second, third and forth methods have been well explained by the books about ultrasonic/piezoelectric motors. Hence only the details of the first method are given here.

Fig. 2.5.1 is used to illustrate the method of generating the traveling waves around a nodal point of non-uniformly excited plate [40, 41]. It uses a tapered metal strip excited by a sandwich type Langevin transducer. Resonance frequency of the actuator is 89.2 kHz, and the operation is around this resonance frequency.

Fig. 2.5.2 shows the out-of-plane vibration displacement distribution of the plate, which indicates that the vibration of the plate root (at $x = 30$ mm) is not uniform along the width direction (or the y direction). Concentric circles with different radii R and centered at a nodal point N ($x = 4$ mm, $y = 0$) were drawn, and the amplitude and phase angle of the out-of-plane vibration displacement on these circles were measured. Fig. 2.5.3 shows the measured results for the circles with $R = 2.1$ mm, 1.4 mm and 0.7 mm. It is seen that the phase of the vibration displacement decreases as position angle θ increases in sections S_1 and S_2. This result means that two local travelling waves around nodal point

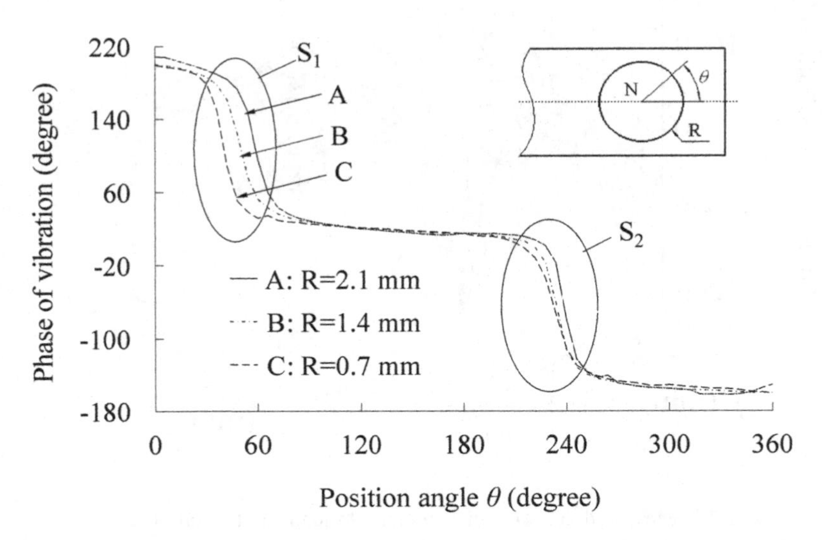

Fig. 2.5.3. Measured distribution of the vibration displacement around the nodal point N.

N are generated in sections S_1 and S_2. Both of them move in the $+\theta$ direction.

Range of the circumferential travelling waves can be extended to a whole circle if a proper vibration excitation structure is used [42, 43]. A rectangular metal plate anti-symmetrically excited by two piezoelectric plates at its root can generate circumferential travelling waves around its nodal points along a whole circle. A round metal plate anti-symmetrically excited by two piezoelectric arcs can also generate circumferential travelling waves around its nodal points along a whole circle. The potential applications of the travelling waves include rotary driving of micro mechanical components and solid particles, and stir in a droplet situated at the nodal points.

2.6 Chladni Effect

Particles on the surface of a flexurally vibrating plate can move to the nodal points, lines and circles (the locations where the vibration displacement is zero) of the plate. This phenomenon was discovered and reported by German physicist Ernst Chladni (1756–1827). In this book, this phenomenon is termed Chladni effect. It not only has applications in the design and construction of musical instruments such as violins and guitars, but also is being used in the ultrasonic actuators such as ultrasonic motors and manipulators. One application of Chladni effect in manipulation technology is in the spin of micro particles [43, 44]. In this application, the micro particles are pushed to and trapped at a nodal point by the Chladni effect, and driven to spin by the circumferential traveling wave around the nodal point.

The impact between the vibrating plate and particles is believed to be the cause for the Chladni effect. Reports on direct experiment proof of this operation principle are very scarce. The following experiments done by the research group of the author of this book indicates that the impact between the vibrating plate and particles is the most likely operation principle. In the experiments, the device shown in Fig. 2.5.1 was used. To observe and characterize the Chladni effect, the shrimp eggs which had a diameter of about 440 μm and density of 680 kg/m^3 were used.

Also the area around nodal point *N*, as shown in Fig. 2.5.1, was used for the experiment.

In one of the experiments, the time it took for a particle initially 1.4 mm away from nodal point *N* to move to the nodal point was measured at a driving voltage of 94.4 $V_{0\text{-}p}$ for the frequency change in 89.05 kHz ~ 89.3 kHz, in normal and vacuum conditions. The ambient pressure was 1 atm and 0.11 atm in the normal and vacuum conditions, respectively, and the ambient temperature in both conditions was 25℃. The measured results are listed in Fig. 2.6.1. It is seen that the Chladni effect still exists in the vacuum operation condition, and moreover the particle's moving speed towards the nodal point increases a little bit in the vacuum. This experiment indicates that the Chladni effect is not likely to be caused by the acoustic streaming and acoustic radiation force, because the acoustic streaming and acoustic radiation force can be ignored under the vacuum condition of 0.11 atm.

In another experiment, the phase angle (relative to the operating voltage) of the vibration displacement around nodal point *N* was measured by a laser Doppler vibrometer (POLYTEC PSV-300F), and the result is shown in Fig. 2.6.2. The phase angle has little change along the

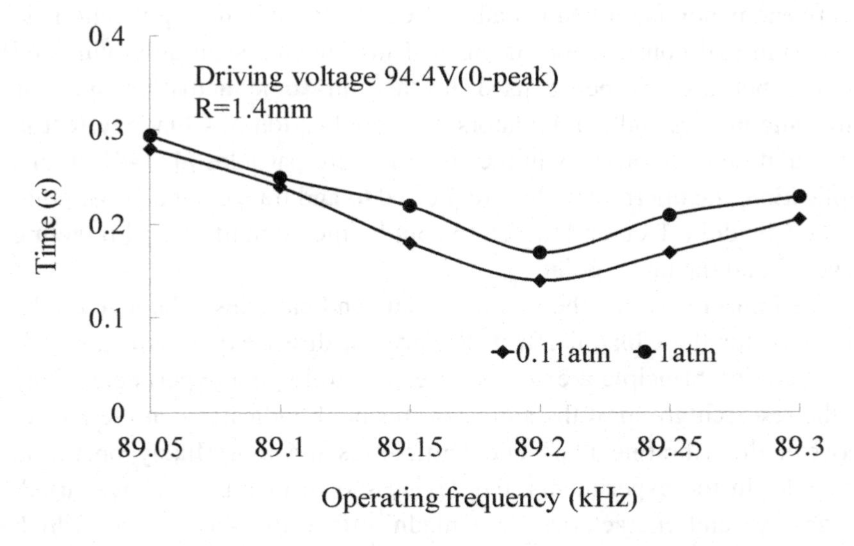

Fig. 2.6.1. A comparison of the Chladni effect in air and vacuum.

radial direction, which indicates that the Chladni effect is not caused by a traveling wave.

It is necessary to mention the inverse Chladni patterns or figures. In the inverse Chladni patterns or figures, microscale particles on a flexurally vibrating plate are pushed to the anti-nodal points, lines or circles (the locations where the vibration displacement is the maximum) of a plate. This phenomenon is caused by the acoustic streaming on the plate surface [45]. Thus it cannot happen in the vacuum operation condition. In this book, this phenomenon is termed inverse Chladni effect.

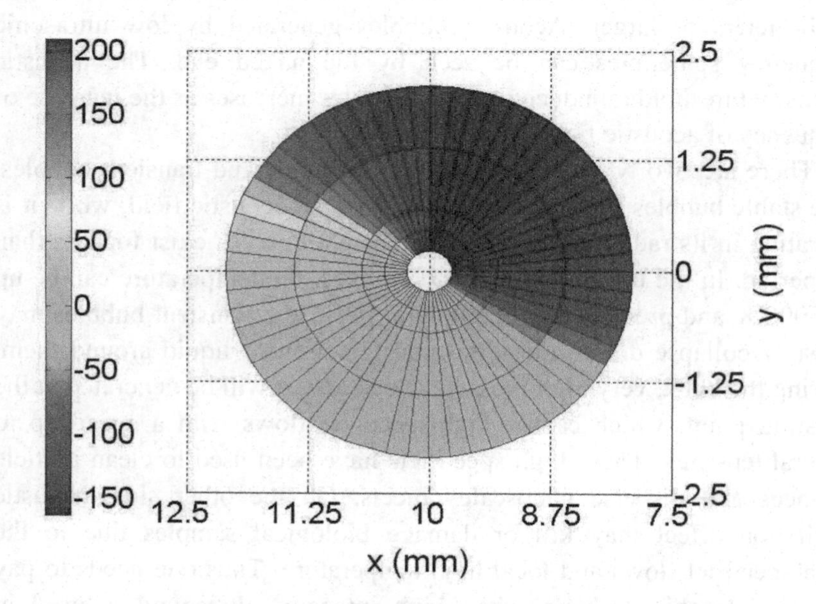

Fig. 2.6.2. Distribution of the phase angle of vibration displacement around the nodal point.

2.7 Acoustic Cavitation and Bjerknes Forces

2.7.1 *Acoustic cavitation*

The acoustic cavitation is a well known phenomenon, in which micro bubbles are generated in liquid with an acoustic field when its intensity of ultrasound is higher than a threshold value [46]. If we ignore the 2nd order pressure, the instantaneous pressure P_T in a sound field is

$$P_T = P_0 + p \tag{2.7.1}$$

where P_0 is the pressure at the field point when there is no sound, and p is the sound pressure. If the amplitude of sound pressure is larger than P_0, P_T can become negative in the negative half cycle of sound pressure. The negative pressure results in voids or cavities in the acoustic field, and these voids or cavities can grow into tiny bubbles. The diameter of acoustically induced bubbles is usually from several microns to several milli-meters or larger. Acoustic bubbles generated by low ultrasonic frequency sometimes can be seen by the naked eye. The acoustic intensity threshold to induce acoustic bubbles increases as the increase of frequency of acoustic field [47].

There are two types of bubbles, i.e. the stable and transient bubbles. The stable bubbles last for many periods of the acoustic field, while it is vibrating in its radial direction. The transient bubbles exist for less than on period. In the ultrasound induced bubbles, the temperature can be up to 5000 K and pressure up to 2000 atm [47]. The transient bubbles may burst or collapse due to the ultrasonic vibration of liquid around them. During the burst, very high pressure shock waves will be generated at the bursting point, which causes high-speed jet flows with a speed up to several tens m/s. These high speed jets have been used to clean particle surfaces and disperse nanoscale objects. On the other side, acoustic cavitation effect may kill or damage biological samples due to the high-speed jet flows and local high temperature. Thus one needs to pay attention to this problem when high intensity ultrasound is used in manipulations.

An oscillation bubble in acoustic cavitation effect can emit sound, and thus generate acoustic radiation force on nearby particles and bubbles. It can also generate microstreaming around itself. The pattern of the microstreaming changes with the oscillation amplitude of the bubble, acoustic medium viscosity and bubble shape [46, 52, 53].

2.7.2 *Bjerknes forces*

A bubble in an acoustic field can be excited to undergo volume pulsation or oscillate in the radial direction of the bubble. Due to the external acoustic field, this bubble experiences an acoustic radiation force. This force is called primary Bjerknes force. In a planar standing-wave field, the primary Bjerknes force on a bubble points to the sound pressure antinode if it is smaller than the resonant size (the size with which the bubble is resonant with the acoustic field), and points to the sound pressure node if it is larger than the resonant size. A detailed mathematical analysis of the primary Bjerknes force can be found in Ref. 48.

For two nearby oscillating bubbles, the sound irradiated from them can generate acoustic radiation force acting on them. This acoustic radiation force is called mutual (or secondary) Bjerknes force. The mutual Bjerknes force makes the two bubbles attract each other if they are both larger or smaller than the resonant size, and repel each other if one is larger than the resonant size and another smaller than the resonant size. More details about the methods of mathematical analysis and experimental measurement of this force can be found in Refs 49-51.

2.8 Sound Induced Intermolecular Force Change

From the acoustic cavitation effect, it is know that when the ultrasonic intensity is greater than the cavitation threshold, acoustic bubbles can be generated in the medium of acoustic field. An interesting question is if the ultrasonic intensity is less than the cavitation threshold, whether there are physical effects other than the acoustic radiation force and acoustic streaming. Based on lots of experimental results, it is deduced that the molecular cohesive force is decreased by ultrasound even if the ultrasonic intensity is less or much less than the cavitation threshold [54-58].

This conclusion is explained as the follows. During the negative half circle of sound pressure, negative sound pressure increases the distance among the molecules of liquid, which is also the reason for acoustic cavitation effect when the acoustic intensity is strong enough. In this case, cohesive force among the molecules such as the hydrogen bong becomes weak. During the positive half circle of sound pressure, positive sound pressure compresses the molecules of liquid. But this cannot increase the cohesive force too much because the molecules repel each other when they are too close. On the time average, ultrasound makes the cohesive force among the molecules of liquid weak. This sound induced change of cohesive force among the molecules of liquid can be expressed by Fig. 2.8.1.

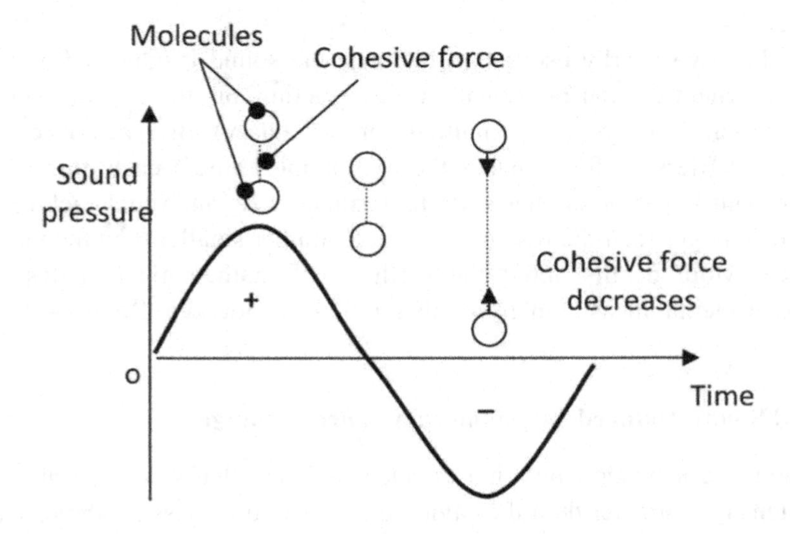

Fig. 2.8.1. A model to explain the effect of ultrasound on intermolecular force.

The weakening of molecular cohesive force in sonicated liquid can cause a decrease of the surface tension of the liquid, and enhance the adsorption of the liquid onto a solid surface. Although the model shown in Fig. 2.8.1 hasn't been investigated quantitatively, the phenomenon has been well used in the transportation of micro fluid, the controlled adsorption, chamber-free extraction and generation of micro droplets.

2.9 Summary and Remarks

Major physical principles and concepts for ultrasonic micro/nano manipulations are given in this chapter, which are very useful in understanding, design and optimization of ultrasonic devices for micro/nano manipulations. To make full use of these principles in ultrasonic micro/nano manipulations, it is necessary to have a complete and quantitative understanding of how manipulation forces in the physical principles change with the working parameters such as operating frequency, vibration amplitude, acoustic field structure, etc. However, at the present stage this understanding is still not sufficient for most of the principles. More computation and modeling work is needed for the ultrasonic manipulation technology.

References

1. Cady, W. C. (1946) *Piezoelectricity*, (McGraw–Hill, New York).
2. Ikeda, T. (1990) *Fundamentals of Piezoelectricity*, (Oxford University Press, Oxford).
3. Yang, J. S. (2005) An Introduction to the Theory of Piezoelectricity, (Springer, New York).
4. Yang, J. S. (2006) *The Mechanics of Piezoelectric Structure*, (World Scientific Publishing Company).
5. APC International Ltd. (2002). *Piezoelectric ceramics: principles and applications*, (Mackeyville, Pennsylvania, USA: APC International Ltd).
6. Ueha, S. and Tomikawa, Y. (1994) *Ultrasonic Motors: Theory and Applications*, (Oxford University Press, Oxford) pp. 32.
7. Zhao, C. S. (2011) *Ultrasonic motors: Technologies and Applications*, (Science Press Beijing & Springer).
8. Morse, P. M. and Ingard, K. U. (1968) *Theoretical Acoustics*, (McGraw–Hill, New York) pp. 241–249, pp. 285–286.
9. Hasegawa, T., Kido, T., Iizuka, T. and Matsuoka, C. (2000). A general theory of Rayleigh and Langevin radiation pressures, 21 (3), (J. Acoust. Soc. Jpn. (E)) pp. 145–152.
10. Liu, Y. Y., Hu, J. H. and Zhao, C. S. (2010). Dependence of acoustic trapping capability on the orientation and shape of particles, IEEE Trans. Ultrason. Ferroelectr. Freq. Control, 57 (6), pp. 1443–1450.

11. Gor'kov, L. P. (1962). On the forces on a small particle in an acoustical field in an ideal fluid, Sov. Phys.–Dokl., 6 (9), pp. 773–775.
12. Liu, Y. Y. and Hu, J. H. (2009). Trapping of particles by the leakage of a standing wave ultrasonic field, J. Appl. Phys., 106 (3), pp. 034903.
13. Hu, J. H., Yang, J. B. and Xu, J. (2004). Ultrasonic trapping of small particles by sharp edges vibrating in a flexural mode, Appl. Phys. Lett., 85 (24), pp. 6042–6044.
14. Hu, J. H., Tay, C., Cai, Y. and Du, J. L. (2005). Controlled rotation of sound–trapped small particles by an acoustic needle, Appl. Phys. Lett., 87 (9), pp. 094104.
15. Hu, J. H., Xu, J., Yang, J. B., Du, J. L., Cai, Y. and Tay, C. (2006). Ultrasonic collection of small particles by a tapered metal strip, IEEE Trans. Ultrason. Ferroelectr. Freq. Control, 53 (3), pp. 571–578.
16. Hu, J. H., Ong, L., Yeo, C. and Liu, Y. Y. (2007). Trapping, transportation and separation of small particles by an acoustic needle, Sens. Actuators, A, 138 (1), pp. 187–193.
17. Yamakoshi, Y. and Noguchi, Y. (1998). Micro particle trapping by opposite phase ultrasonic traveling waves, Ultrasonics, 36 (8), pp. 873–878.
18. Miller, D. L. and Nyborg, W. L. (1979). Platelet aggregation induced by ultrasound under specialized conditions in vitro, Science, 205 (4405), pp. 505–507.
19. Coakley, W. T., Bardsley, D.W. and Grundly, M.A. (1989). Cell manipulation in ultrasonic standing wave fields, J. Chem. Tech. Biotechnol., 44 (1), pp. 43–62.
20. Takeuchi, M. and Yamanouchi, K. (1994). Ultrasonic micromanipulation of small particles in liquid, Jpn. J. Appl. Phys., 33, pp. 3045–3047.
21. Benes, E., Groschl, M., Nowotny, H., Trampler, F., Keijzer, T., Bohm, H., Radel, S., Gherardini, L., Hawkes, J.J., Konig, R. and Delouvroy, C. (2001). Ultrasonic separation of suspended particles, IEEE Ultrason. Symp., pp. 649–659
22. Wiklund, M., Nilsson, S. and Hertz, H.M. (2001). Ultrasonic trapping in capillaries for trace–amount biomedical analysis, J. Appl. Phys., 90 (1), pp. 421–426.
23. Hu, J. H. and Santoso, A.K. (2004). A Pi–shaped ultrasonic tweezers concept for manipulation of small particles, IEEE Trans. Ultrason. Ferroelectr. Freq. Control, 51 (11), pp. 1499–1507.
24. Haake, A. and Dual, J. (2005). Contactless micromanipulation of small particles by an ultrasound field excited by a vibrating body, J. Acoust. Soc. Am., 117 (5), pp. 2752–2760.
25. Wu, J. (1991). Acoustic tweezers, J. Acoust. Soc. Am., 89 (5), pp. 2140–2143.
26. Mitri, F.G. (2009). Langevin acoustic radiation force of a high–order bessel beam on a rigid sphere, IEEE Trans. Ultrason. Ferroelectr. Freq. Control, 56 (5), pp. 1059–1064.
27. Coakley, W. T., and Nyborg, W. L. (1978). Cavitation; dynamics of gas bubbles; applications. *Ultrasound: Its applications in medicine and biology*, 3, pp. 77–159 (in New York).
28. Young, F. R. (1999) *Cavitation*, (Imperial College Press, London) pp. 77 & 161.

29. Chung, S. K., Rhee, K. and Cho, S. K. (2010). Bubble actuation by electrowetting–on–dielectric (ewod) and its applications: a review, International Journal of Precision Engineering and Manufacturing, 11 (6), pp. 991–1006.
30. Friend, J. and Yeo, L.Y. (2011). Microscale acoustofluidics: microfluidics driven via acoustics and ultrasonics, Rev. Mod. Phys., 83, pp. 647–704.
31. O'Rorke, R.D., Wood, C.D., Walti, C., Evans, S.D., Davies, A.G. and Cunningham, J.E. (2012). Acousto–microfluidics: transporting microbubble and microparticle arrays in acoustic traps using surface acoustic waves, J. Appl. Phys., 111, 94911.
32. Abramov, O. V. (1998) *High–Intensity Ultrasonics*, (Gordon and Breach Science Publishers, Singapore) pp. 124–139.
33. Nyborg, W. L. (1966) *Physical Acoustics* (edited by W. P. Mason and R. N. Thurston) (Academic Press, New York) pp. 267–303.
34. Li, N., Hu, J., Li, H., Bhuyan, S. and Zhou, Y. (2012). Mobile acoustic streaming based trapping and 3–dimensional transfer of a single nanowire, Appl. Phys. Lett., 101 (9), 93113.
35. Zhou, Y.J., Hu, J.H. and Bhuyan, S. (2013). Manipulations of silver nanowires in a droplet on a low–frequency ultrasonic stage, IEEE Trans. Ultrason. Ferroelectr. Freq. Control, 60 (3), pp. 622–629.
36. Lighthill, J. (1978) *Waves in Fluids*, (Cambridge University Press, Cambridge) pp. 329, pp. 344–350.
37. Hu, J.H., Nakamura, K. and Ueha, S. (1996). Optimum operation conditions of an ultrasonic motor driving fluid directly, Jpn J. Appl. Phys., 35, pp. 3289–3294.
38. Lee, C.P. and Wang, T.G. (1989). Near–boundary streaming around a small sphere due to two orthogonal standing waves, J. Acoust. Soc. Am., 85 (3), pp. 1081–1088.
39. Tomikawa, Y. (1998) *Vibration Theory of Ultrasonic Electronics*, (Asakura–shoten, Tokyo) pp. 241–248 (in Japanese).
40. Zhang, X., Zheng, Y. and Hu, J. (2008). Sound controlled rotation of a cluster of small particles on an ultrasonically vibrating metal strip, Appl. Phys. Lett., 92, 024109.
41. Hu, J. H., Zhu, X. B., Zhou, Y. J. and Li, N. (2012). Principle of the rotation of small particles around a nodal point of strip in flexural vibration, Sens. Actuators, A: Physical, 178, pp. 202–208.
42. Li, H. Q., Hu, J. H., Zhu, X. B. and Zhou, Y. J. (2011). Drive of micro particles around vibration node of copper substrate vibration in flexural mode, Proceedings of the 2011 Symposium on Piezoelectricity, Acoustic Waves and Device Applications (SPAWDA11), IEEE, pp. 44–47.
43. Zhou, Y. J., Li, H. Q. and Hu, J. H. (2012). An ultrasonic stage for controlled spin of micro articles, Rev. Sci. Instrum., 83, 045004.
44. Zhu, X. B. and Hu, J. H. (2013). Ultrasonic drive of small mechanical components on a tapered metal strip, Ultrasonics, 53, pp. 417–422.

45. Dorrestijn, M., Bietsch, A., Acikahn, T., Raman. A., Hegner. H., Meyer, E. and Gerber, Ch. (2007). Chladni figures revisited based on nanomechanics, *Phys. Rev. Lett.*, 98, 026102.
46. Ronald–Young, F. (1999) *Cavitation*, (Imperial College Press, London) pp. 6 &160.
47. Mason, T. J. and Lorimer, J. P. (2002) *Applied Sonochemistry: Use of Power Ultrasound in Chemistry and Processing*, (Wiley–VCH Verlag GmbH & Co. KGaA, Weiheim, Germany), pp. 40 & 45.
48. Leighton, T. G., Walton, A. J. and Pickworth, M. J. (1990). Primary bjerknes forces, Eur. J. Phys., 11, pp. 47–50.
49. Weiser, M. A. H., Apfel, R. E. and Neppiras, E. A. (1984). Interparticle forces on red cells in a standing wave field, Acoustica, 56 (2), 114–119.
50. Mettin, R., Akhatov, I., Parlitz, U., Ohl, C. D. and Lauterborn, W. (1997). Bjerknes forces between small cavitation bubbles in a strong acoustic field, Phys. Rev. E: Stat. Phys., Plasmas, Fluids, 56, pp. 2924–2931.
51. Crum, L. A. (1975). Bjerknes forces on bubbles in a stationary sound field, J. Acoust. Soc. Am., 57, pp. 1363–1370.
52. Rozenberg, L. D. (1973) *Physical Principles of Ultrasonic Technology*, (Plenum Press, New York, USA).
53. Elder, S. A. (1959). Cavitation microstreaming, J. Acoust. Soc. Am., 31, pp.54–64.
54. Hu, J. H., Tan, C. L. and Hu, W. Y. (2007). Ultrasonic microfluidic transportation based on a twisted bundle of thin metal wires, Sens. Actuators, A: Physical, 135, pp. 811–817.
55. Tan, Z., Teo, G. (2008). Ultrasonic generation and rotation of a small droplet at the tip of a hypodermic needle, J. Appl. Phys., 104, 104902.
56. Hu, J. H., Li, N. and Zhou, J. J. (2011). Controlled adsorption of droplets onto anti–nodes of an ultrasonically vibrating needle, J. Appl. Phys., 110, 054901.
57. Yun, C., Hasegawa, T., Nakamura, K and Ueha, S. (2004). An ultrasonic suction pump with no physically moving parts, Jpn. J. Appl. Phys., 43, pp. 2854–2868.
58. Bhuyan, S. (2010). *Wireless drive of Piezoelectric Components*, Chapter 7 (PhD thesis, Nanyang Technological University, Singapore) pp. 147.
59. Nyborg, W. L. (1958). Acoustic Streaming near a Boundary, J. Acoust. Soc. Am., 30, pp. 329–339.
60. Lee, C. P. and Wang, T. G. (1989). Near–boundary streaming around a small sphere due to two orthogonal standing waves, J. Acoust. Soc. Am., 85, pp. 1081–1088.
61. Hu, J. H., Nakamura, K. and Ueha, S. (1997). An analysis of a noncontact ultrasonic motor with an ultrasonically levitated rotor, Ultrasonics, 35, pp.459–467.
62. Hashimoto, Y., Koike, Y. and Ueha, S. (1998). Transporting objects without contact using flexural traveling waves, J. Acoust. Soc. Am., 103, pp. 3230–3233.
63. Hu, J. H., Nakamura K. and Ueha, S. (1999). A noncontact ultrasonic motor with the rotor levitated by axial acoustic viscous force, *Electronics and Communications in Japan (Part III)*, 82 (4), pp. 56–63. John Wiley & Sons, Inc. USA.

64. Ueha, S., Hashimoto, Y. and Koike, Y. (2000). Non–contact transportation using near–field acoustic levitation, Ultrasonics, 38, pp. 26–32.
65. Hu, J. H., Cha, K. C. and Lim, K. C. (2004). New Type of Linear Ultrasonic Actuator Based on a Plate–Shaped Vibrator with Triangular Grooves, IEEE Trans. Ultrason. Ferroelectr. Freq. Control, 51, pp. 1206–1208.
66. Hu, J. H., Li, G. R., Chan, H. L. W. and Choy, C. L. (2001). A standing wave type noncontact linear ultrasonic motor. IEEE Trans. Ultrason. Ferroelectr. Freq. Control, 48, pp. 699–708.
67. Hu, J. H., Nakamura, K. and Ueha, S. (2003). Stability analysis of an acoustically levitated disk, IEEE Trans. Ultrason. Ferroelectr. Freq. Control, 50, pp.117–127.

Chapter 3

Ultrasonic Trapping of Micro Solids

In this chapter, examples of contact and noncontact type trapping of micro solids are given. In each example, the structure and principle of the device are given first, and then manipulation characteristics are given and discussed. Sections 3.1-3.3 show examples of the contact type trapping, and Section 3.4 shows a π-shaped ultrasonic tweezers which may operate in a noncontact means. In the contact type trapping, the ultrasonic radiation point, line and face are employed to generate the acoustic radiation force. And in noncontact type trapping of the π-shaped ultrasonic tweezers, a standing wave field inside the tweezers is employed.

3.1 On a Radiation Point

In this example, trapping, transportation and separation of small particles in liquid by an acoustic needle are demonstrated [1-3]. After giving the structure and analyzing the mechanism of trapping small particles by the tip of the acoustic needle, characteristics of trapping, transportation and separation of tiny particles such as Flying Color seeds, grass seeds and shrimp eggs are presented. It provides a method of trapping, transporting and separating small particles, which may have potential applications in bioengineering and other areas.

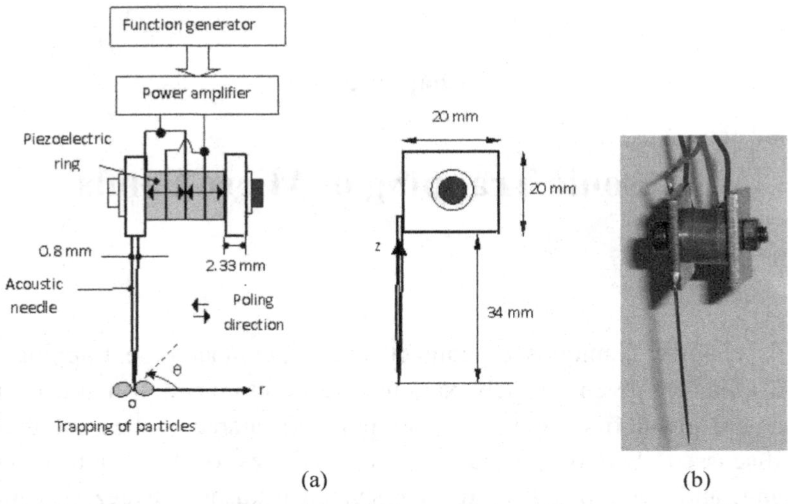

Fig. 3.1.1. The ultrasonic transducer with a metal needle to trap small particles. (a) Structure and size. (b) Photo.

3.1.1 *Construction and principle*

Fig. 3.1.1 shows the ultrasonic transducer used in our experiments. The acoustic needle made of stainless steel is welded onto the side of one of the two rectangular stainless steel plates in a sandwich type piezoelectric transducer. A multilayer piezoelectric vibrator consisting of four piezoelectric rings is pressed by the two plates via a bolt structure. The neighboring piezoelectric rings aligned with opposite poling directions. The size of each stainless plate is $20\times20\times2.3$ mm^3, and outer diameter, inner diameter and thickness of each piezoelectric ring are 20 mm, 12 mm and 2.4 mm, respectively. The electromechanical quality factor Q_m, piezoelectric coefficient d_{33}, and relative dielectric constant $\varepsilon_{33}^T / \varepsilon_0$ of the piezoelectric rings are 2000, 325×10^{-12} m/v and 1450, respectively. The needle has a length of 34 mm from the edge of the stainless steel plate, and diameter of 0.8 mm which gradually decreases down to its tip from z = 11 mm. When an AC voltage with a frequency near the resonance frequency of the thickness vibration mode of the ultrasonic transducer is applied to the transducer, a thickness vibration can be excited in the transducer. According to our measurement by impedance analyzer 4194A (Hewlett Packard), the resonance frequency of the first and second order thickness vibration mode is about 66.75 kHz and 131 kHz, respectively. Due to the arrangement of the piezoelectric rings shown in Fig. 3.1.1, the first order thickness mode vibration is much stronger than that the second one. In the following experiments, the first order thickness vibration mode is used. The weaker second order thickness vibration mode may be used in the rotation of small particles in which a strong acoustic streaming is not desired.

When the transducer operates near the first order thickness vibration mode and the driving condition is proper, small particles such as Flying Color seeds, grass seeds and shrimps eggs (Horti-Hora Ltd, Singapore) near the tip of the vibrating needle in water can be sucked to and trapped by the tip. Fig. 3.1.2 shows the trapping of Flying Color seeds in water. The particles used in our experiments can sink to the vessel bottom because they have absorbed sufficient water.

Fig. 3.1.2. Flying Color seeds trapped by the tip of the vibrating needle in water.

Due to the thickness vibration mode, it can be expected that the needle vibrates back and forth in the *rz* plane. The sound pressure along the two lines at $r = 0.5$ mm, $\theta = 0$ and $r = 0.5$ mm, $\theta = 90°$ was measured by a 1 mm needle hydrophone (SN945, Precision Acoustics, UK) in the range from $z = 0$ to $z = 25$ mm in water, when an voltage with a frequency of 66.75 kHz and amplitude of 15 $V_{p\text{-}p}$ was applied to the transducer. The acoustic needle submerged in water was perpendicular to the needle hydrophone probe in the measurement. Fig. 3.1.3 shows the sound pressure versus z at $\theta = 0°$ and 90°. In the measurement, the acoustic needle length under water was 29 mm and the distance from the hydrophone tip to acoustic needle surface was 0.5 mm. From Fig. 3.1.3, it is seen that the sound pressure is larger at $\theta = 0$ than at $\theta = 90°$. This indicates that the acoustic needle vibrates at a flexural mode in the rz plane.

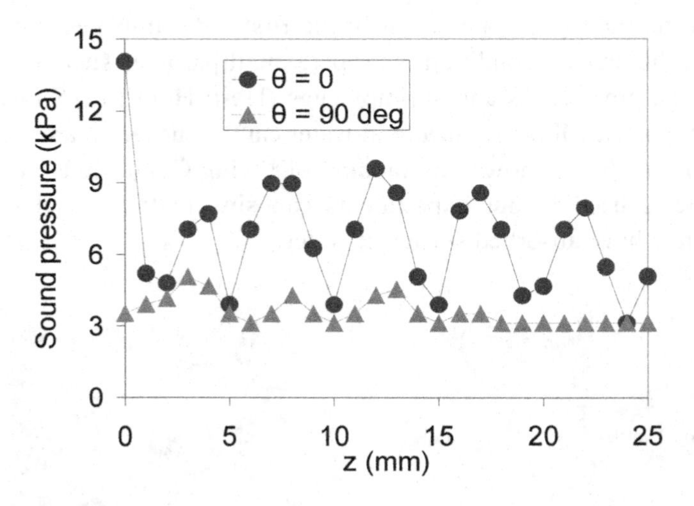

Fig. 3.1.3. Sound pressure distribution along the length direction of the vibrating needle. The sound pressure is in peak-peak value.

The 3D finite element method (FEM) calculation, which employs the acoustic radiation force theory given in Section 2.3, shows that micro particles near the acoustic needle tip experience an acoustic radiation force pointing to the tip. This force pushes the micro particles onto the acoustic needle tip, and traps them at the tip by a friction force between the particles and acoustic needle tip. Based on the calculation, when the acoustic needle root has vibration amplitude of 1 μm (0-peak), the acoustic radiation force is of an order of magnitude of 10 μN for a spherical particle with 0.5 mm radius in the vicinity of the acoustic needle tip in water.

3.1.2 *Experimental results and discussion*

To measure the characteristics of trapping, transportation and separation, plastic vessels filled with water or other liquid were used, and the height of the liquid in the vessels is 24 mm. The diameter of the vessels is much larger than the wavelength of sound wave in the liquid. The tip of the acoustic needle was approximately located in the center region of the sound field. The measured sound field versus position r along the line $\theta = 0$, $z = 0$ is showed in Fig. 3.1.4. In the measurement, the operating frequency f and driving voltage are 66.75 kHz and 15 V_{p-p}, respectively. Fig. 3.1.4 shows that the sound field not very close to the tip is a traveling wave because no nodal point is detected at the location (r = sound speed in water / frequency */4* = 1500 m/s / 66.75 kHz / 4 = 5.6 mm, or 3×5.6 mm = 16.8 mm) where the sound pressure would be minimum if it were a standing wave sound field. This is because of a large acoustic damping of plastic wall and large distance between the tip and the plastic wall.

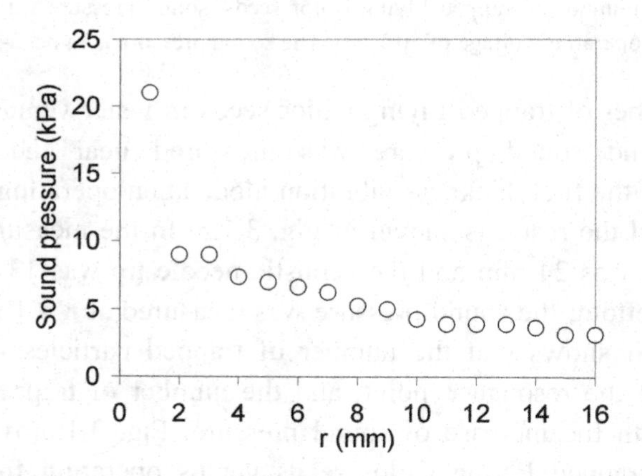

Fig. 3.1.4. Sound pressure distribution along position r at $\theta = 0$ and $z = 0$. The sound pressure is in peak-peak value.

Flying Color seeds, grass seeds and shrimp eggs were used in the experiments. The average radius of Flying Color seeds, grass seeds and shrimp eggs is 0.54 mm, 0.56 mm and 0.12 mm, respectively. The density of dry Flying Color seeds, grass seeds and shrimp eggs is 0.93 g/cm^3, 0.53 g/cm^3 and 0.55 g/cm^3, respectively.

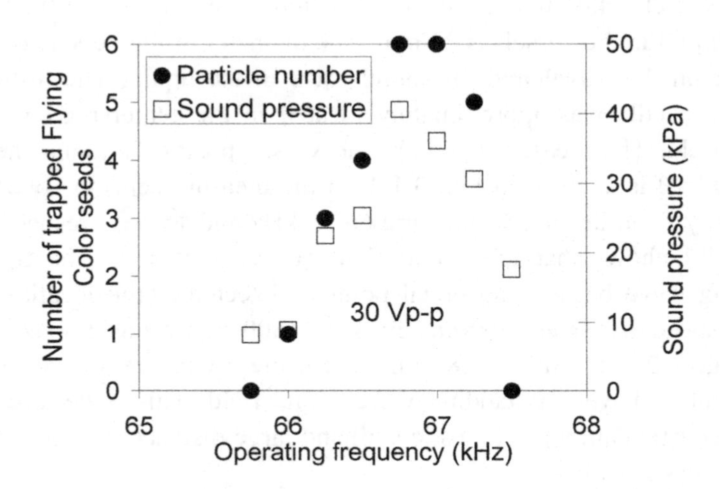

Fig. 3.1.5. The number of trapped Flying Color seeds, sound pressure versus operating frequency at an operating voltage of 30 Vp-p. The sound pressure is in peak-peak value.

The number of trapped Flying Color seeds in water versus operating frequency and sound pressure was measured near the resonance frequency of the first thickness vibration mode at an operating voltage of 30 Vp-p, and the result is shown in Fig. 3.1.5. In the measurement; the water height was 24 mm and the acoustic needle tip was 13 mm above the vessel bottom; the sound pressure was measured at $r = 1$ mm, $\theta = 0$, and $z = 0$. It shows that the number of trapped particles reaches the maximum at the resonance point, and the number of trapped particles increases with the increase of sound pressure. Fig. 3.1.6(a) shows the number of trapped Flying Color seeds versus operating frequency at driving voltages of 7.5 Vp-p, 15 Vp-p and 50 Vp-p. As the driving voltage increases, the maximum number of trapped particles increases. Fig. 3.1.6(b) shows the measured maximum number of trapped Flying Color seeds versus driving voltage. As the driving voltage increases, the maximum number of trapped particles increases and tends to saturate when the voltage is too large.

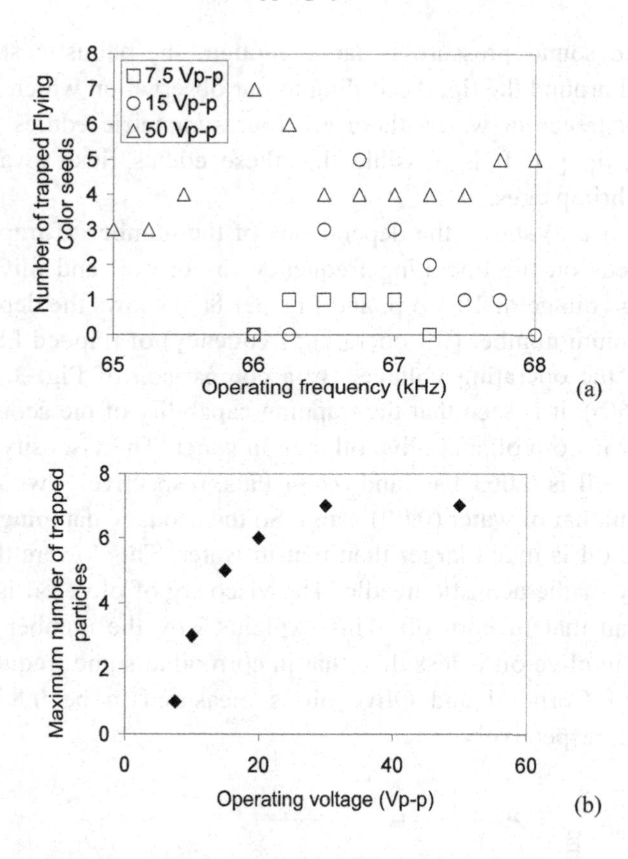

Fig. 3.1.6. The number of trapped particles at different voltages. (a) The number of trapped Flying Color seeds versus operating frequency at operating voltages of 7.5 Vp-p, 15 Vp-p and 50 Vp-p. (b) The maximum number of trapped Flying Color seeds versus operating voltage.

The number of trapped shrimp eggs in water versus sound pressure was measured and the result is shown in Fig. 3.1.7. In the measurement; the water height was 24 mm and the acoustic needle tip was 10 mm above the bottom of vessel; operating voltage was 30 Vp-p; sound pressure was measured at $r = 1$ mm, $\theta = 0$, and $z = 0$. It shows that the number of trapped particles increases as the sound pressure increases, and tends to saturate when the sound pressure exceeds a certain value.

When the sound pressure is large enough, the acoustic streaming is generated around the tip. According to our observation which is based on black ink tracer in water, there are four symmetric eddies around the vibrating tip [1]. It is possible that these eddies flush away some of trapped shrimp eggs.

Fig. 3.1.8(a) shows the dependence of the number of trapped Flying Color seeds on the operating frequency in corn oil and olive oil at an operating voltage of 10 Vp-p, and Fig. 3.1.8(b) shows the dependence of the maximum number (for operating frequency) of trapped Flying Color seeds on the operating voltage. By a comparison of Fig. 3.1.8(b) with Fig. 3.1.6(b), it is seen that the trapping capability of the acoustic needle is weaker in corn oil and olive oil than in water. The viscosity of corn oil and olive oil is 0.063 Pa·s and 0.084 Pa·s, respectively, which is much larger than that of water (0.001 Pa·s). So the acoustic damping in corn oil and olive oil is much larger than that in water. This lowers the trapping capability of the acoustic needle. The viscosity of olive oil is a little bit larger than that of corn oil. This explains why the number of trapped particles in olive oil is less than that in corn oil at some frequencies. The density of Corn oil and Olive oil is measured to be 0.87 g/ml and 0.84 g/ml, respectively.

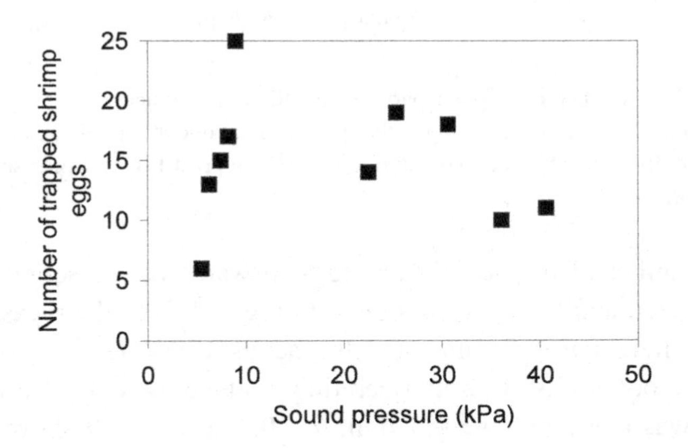

Fig. 3.1.7. The number of trapped shrimp eggs versus sound pressure. The sound pressure is in peak-peak value.

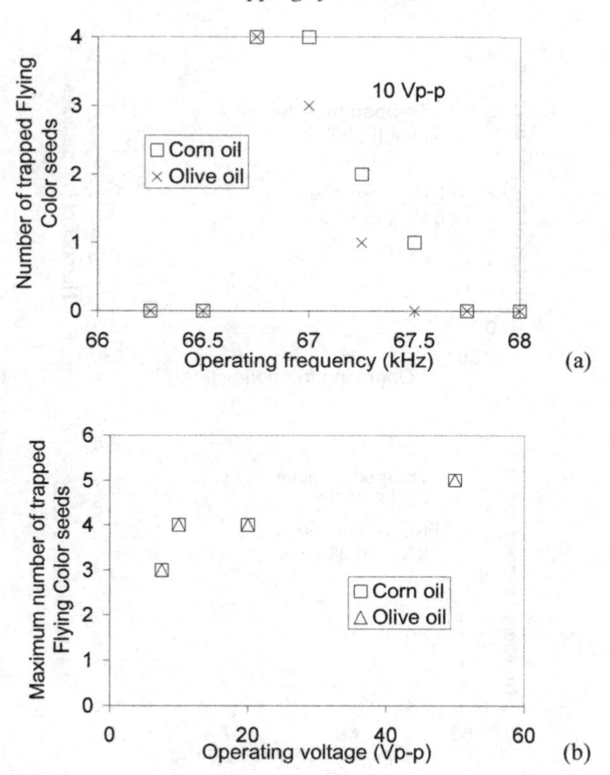

Fig. 3.1.8. Trapping of particles in Corn oil and Olive oil. (a) The number of trapped Flying Color seeds versus operating frequency at an operating voltage of 10 Vp-p. (b) The maximum number of trapped Flying Color seeds for operating frequency versus operating voltage.

The trapped particles can be transported from one location to another in water. Fig. 3.1.9 shows the number of trapped particles at the initial position, and the number of lost particles during the transportation versus operating frequency at an average transportation speed of 0.88 cm/s, transportation distance of 17.5 cm and different voltages. In the experiment, the water height in the vessel was 2.4 cm; the distance from the acoustic needle tip to the vessel bottom was 0.5 cm; the transportation direction was parallel to the vibration direction of the acoustic needle; Flying Color seeds were used. Fig. 3.1.9 shows that the

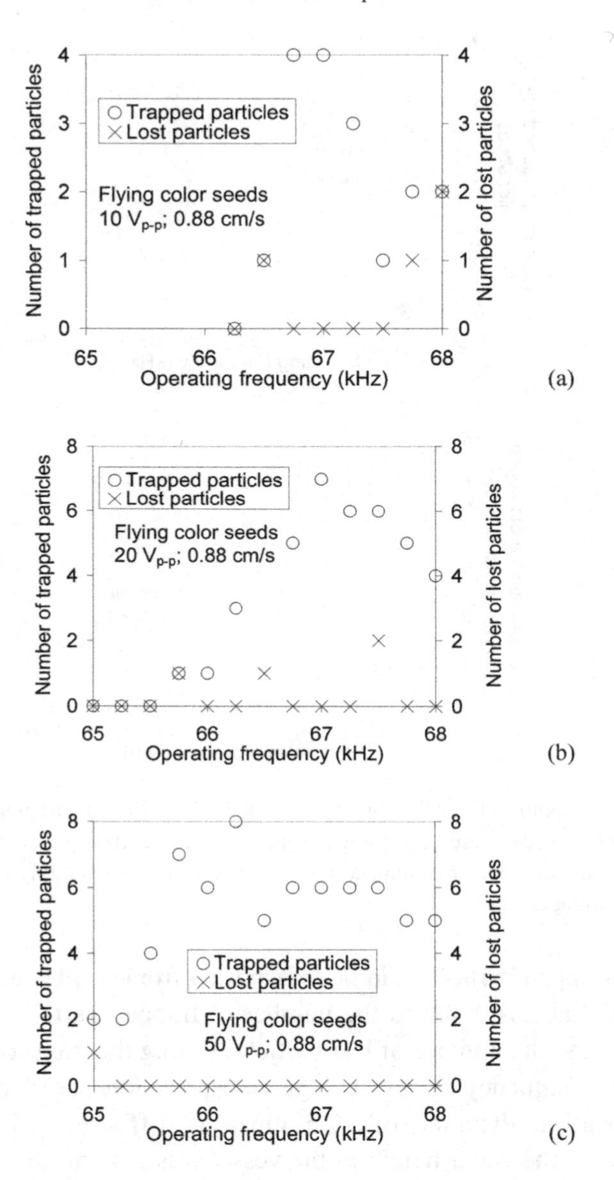

Fig. 3.1.9. The numbers of trapped Flying Color seeds and lost Flying Color seeds during the transportation versus operating frequency at operating voltages of (a) 10 Vp-p, (b) 20 Vp-p and (c) 50 Vp-p. The average transportation speed is 0.88 cm/s.

trapped particles can be transported in water by the acoustic needle. It also shows that there exists a frequency range near the resonance point in which there is no particle lost and this frequency range becomes wider when the operating voltage increases. These phenomena are because the sound pressure is relatively large near a resonance point and it increases when the operating voltage increases near a resonance point. Fig. 3.1.10 shows the frequency dependence of the number of lost Flying Color seeds when the transportation speed is 1.75 cm/s and 3.5 cm/s, and operating voltage is 50 Vp-p. The other measurement conditions are the same as those for Fig. 3.1.9. From the figure, it is known that the particle lose during the transportation may be avoided even at a relatively high transportation speed if a strong sound field is used.

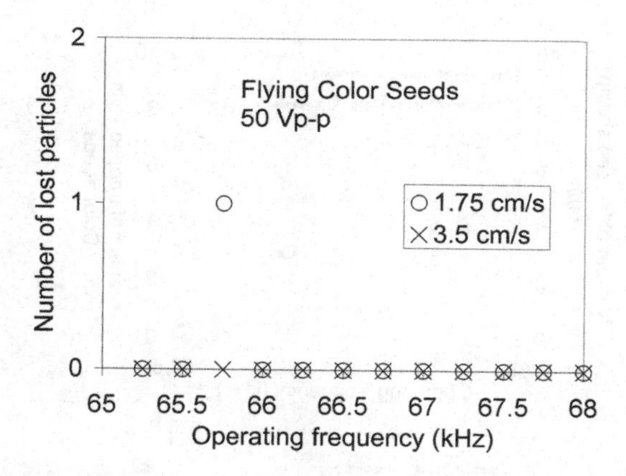

Fig. 3.1.10. The number of lost Flying Color seeds during the transportation versus operating frequency at the average speed of 1.75 cm/s and 3.5 cm/s, and an operating voltage of 50 Vp-p.

Fig. 3.1.11 shows the number of trapped grass seeds at the initial position, and the number of lost particles during the transportation versus operating frequency at an operating voltage of 7.5 Vp-p. The other measurement conditions are the same as those for Fig. 3.1.9. Although their relatively large size, there also exits a frequency range in which there is no particle lost during the transportation.

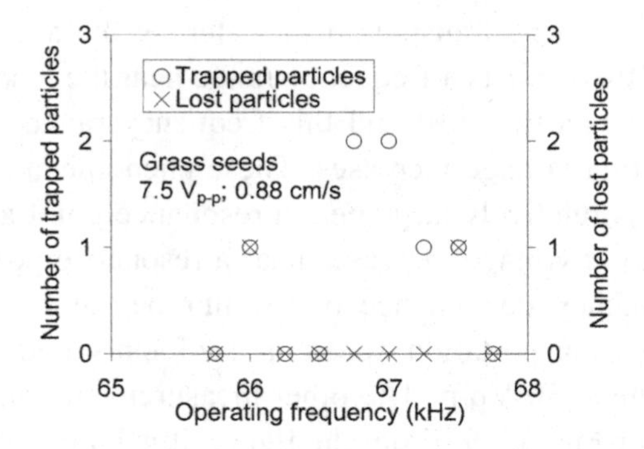

Fig. 3.1.11. The numbers of trapped grass seeds and lost Flying Color seeds during the transportation versus operating frequency at an operating voltage of 7.5 Vp-p and transportation speed of 0.88 cm/s.

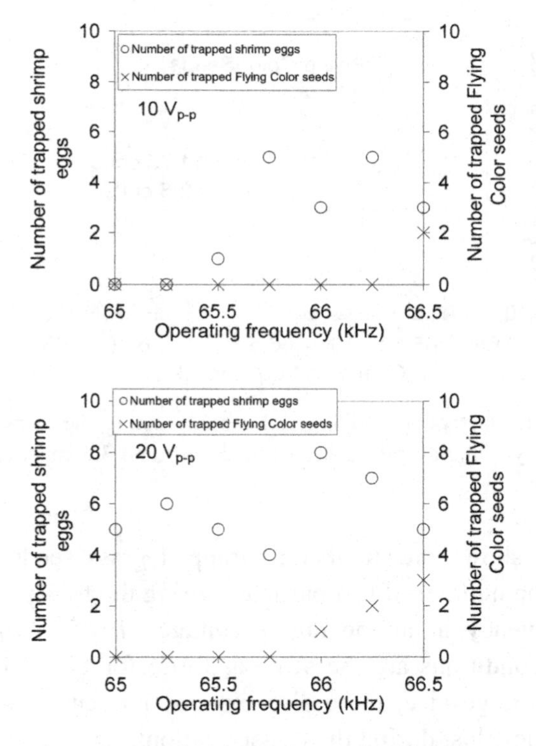

Fig. 3.1.12. The numbers of trapped Flying Color seeds and shrimp eggs versus operating frequency at operating voltages of (a) 10 Vp-p and (b) 20 Vp-p.

In the experiments of investigating the possibility to separate different particles by the acoustic needle, the tip of the vibrating needle was inserted into the mixture of shrimp eggs and Flying Color seeds on the bottom of the vessel with water, and then lifted to a 5 mm height from the bottom for observing whether both of the particles were trapped or not. Fig. 3.1.12 shows the measured numbers of trapped shrimp eggs and Flying Color seeds versus operating frequency at operating voltages of 10 Vp-p and 20 Vp-p. From the figure, it is seen that there exist some frequency ranges in which only shrimp eggs can be trapped and Flying Color seeds cannot. Due to the larger density of Flying Color seeds, the minimum sound pressure to trap Flying Color seeds must be larger than that to trap shrimp eggs. This is why a frequency range in which only shrimp eggs are trapped can be found.

3.1.3 *Summary*

The acoustic needle can trap, transport and separate micro particles such as Flying Color seeds, grass seeds and shrimp eggs in liquid by its ultrasonically vibrating tip. The trapping capability increases with the increase of sound pressure around the tip, and becomes saturated when the sound pressure is too large. It may also be affected by the viscosity of the acoustic medium. And a large viscosity reduces the trapping capability. The trapped particles can be transported in water by moving the needle and the particle loss during the transportation may be prevented by using a strong enough vibration. The acoustic needle can also separate different particles in water by the difference in their densities. Therefore, this work provides a new method of trapping, transporting and separating small particles, which has potential applications in bioengineering and other areas.

3.2 On a Radiation Line

Trapping of micro particles by two parallel ultrasonic radiation lines in air is demonstrated in this section [4, 5]. The contents of this section include the device structure, principle analysis, trapping characteristics and discussion.

3.2.1 *Structure and principle analysis*

The structure of the transducer used to trap particles in our experiment is shown in Fig. 3.2.1. In this transducer, two identical metal strips are clamped to a Langevin transducer, as shown in Fig. 3.2.1(a). The metal strips made of aluminum have the shape and size shown in Fig. 3.2.1(b) and (c). The upper part is a rectangular metal plate, and the lower part a V-shaped metal strip. The upper part has a size of 40 mm × 45 mm × 1.5 mm with a 10 mm diameter hole at its center; the lower part has a length of 99 mm, width of 22.5 mm and thickness of 1.5 mm, and tapers off from the upper end to the lower end. In this way, a triangular air gap is formed between the two V-shaped metal strips, which has a thickness h_a of 1.5 mm at the end, as shown in Fig. 3.2.1(c). The Langevin transducer has a resonance frequency of 25.3 kHz.

When an AC voltage with a frequency close to the resonance frequency of the ultrasonic transducer is applied, a flexural vibration is excited in the metal plates. This flexural vibration will generate a sound field, and the sound field near the lower end of gap can generate an acoustic radiation force to suck the particles to the lower end of strips.

Acoustic radiation force $\boldsymbol{F}$ on a rigid immovable object in a sound field is given by the following integration over the surface of the object.

$$\boldsymbol{F} = \left\langle \iint_S (K - P)\boldsymbol{n} dS \right\rangle \tag{3.2.1}$$

where the notation $\langle\ \rangle$ denotes time average over one period, K is the kinetic energy density, P is the potential energy density, and $\boldsymbol{n}$ is the outward normal unit vector of the surface.

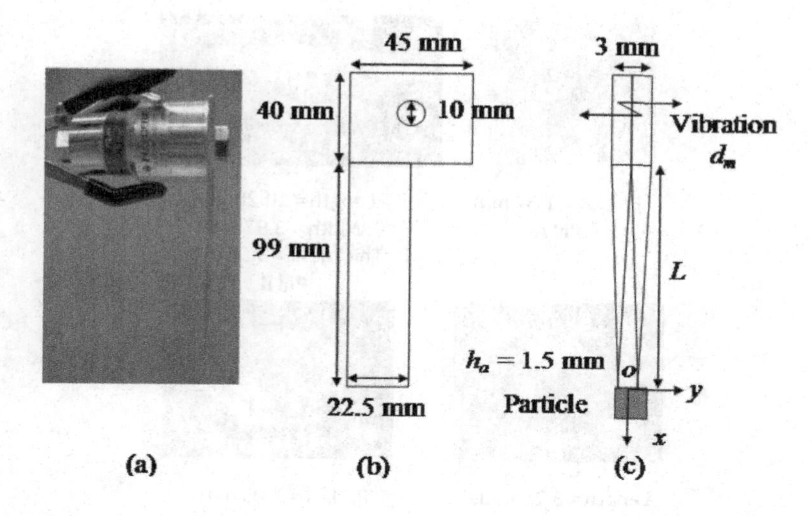

Fig. 3.2.1. Structure and size of the ultrasonic transducer with two V-shaped metal strips. (a) Structure of the ultrasonic transducer. (b) Shape and size of the metal strip. (c) Air gap formed by the two V-shaped metal strips.

To understand the operating principle, the acoustic radiation force on a cube rigid particle of 3 mm×3 mm×3 mm at the inlet of the air gap is calculated by COMSOL Multiphysics software. According to the FEM calculation, it is known that $\iint_S \langle K \rangle dS >> \iint_S \langle U \rangle dS$ on the top surface of the cube particle, and $\iint_S \langle K \rangle dS$ and $\iint_S \langle U \rangle dS$ on the side and bottom surfaces of the particle are less than 1% of that on the top surface, which means that they are negligible. So $\boldsymbol{F}$ has the same direction as $\boldsymbol{n}$ vector of the top surface, and the particle is sucked to the sharp edge of the strips.

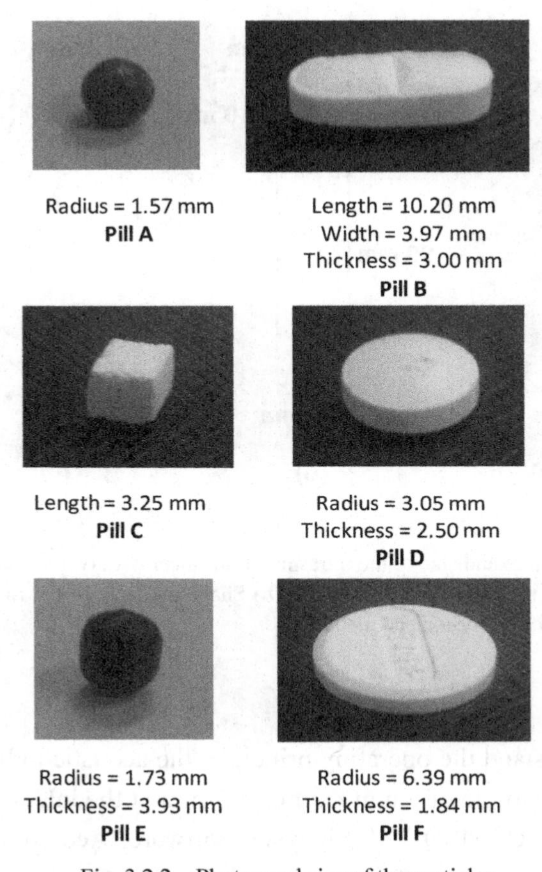

Fig. 3.2.2. Photos and size of the particles.

3.2.2 *Results and discussion*

Particles used in the experiment are medicine pills. The size and shape of the particles are shown in Fig. 3.2.2, and the mass, volume and density of them are shown in Table 3.2.1. The experimental procedure for trapping particles is shown in Fig. 3.2.3. The lower end of the vibrating metal strips is inserted into a collection of particles in a container, and then the transducer is lifted up. It was observed that all kinds of pills shown in Fig. 3.2.2 could be sucked to and trapped on the lower end of the metal strips in air, as shown in Fig. 3.2.4.

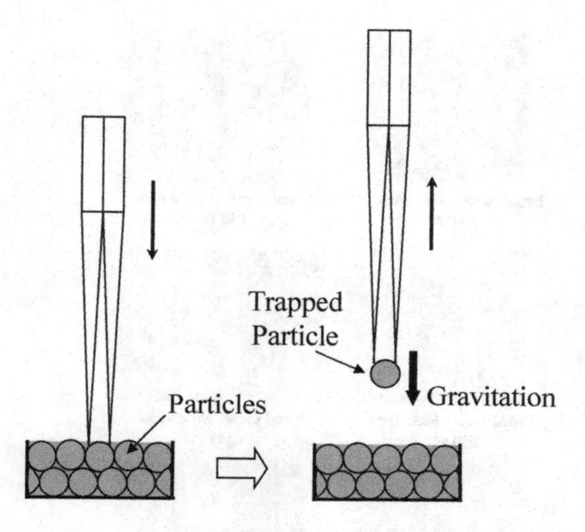

Fig. 3.2.3. Experimental procedure for trapping particles.

Table 3.2.1. Mass, volume and density of the particles.

Pill Type	Mass (g)	Volume (cm^3)	Density (g/cm^3)
Pill A	0.0194	0.0162	1.1975
Pill B	0.1200	0.1215	0.9877
Pill C	0.0400	0.0343	1.1662
Pill D	0.0784	0.0731	1.0725
Pill E	0.0316	0.0370	0.8541
Pill F	0.2560	0.2360	1.0847

Figure 3.2.5 shows the experimental relationship between the number of trapped pills *A* and driving frequency at different input voltages in air. It can be observed that the number of trapped pills reaches a maximum at a particular driving frequency or in a particular driving frequency range for a given voltage, which is caused by the resonance of the transducer system.

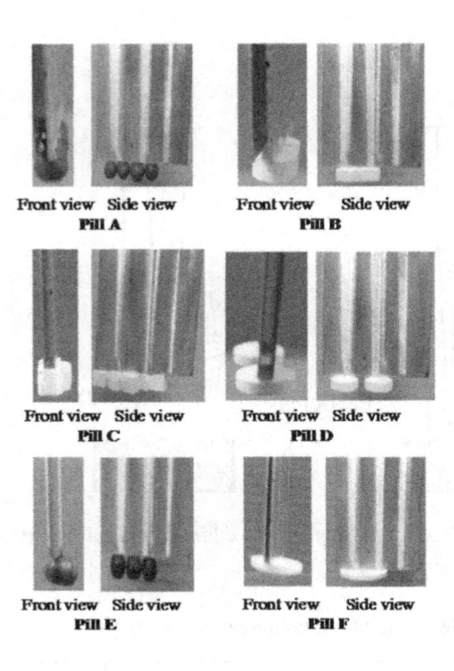

Fig. 3.2.4. Photos of trapped particles.

Figure 3.2.6 shows a measured relationship between the y-direction vibration displacement amplitude (0-peak) at the tip of the V-shaped metal strip and the number of trapped particle for different particle types. From Fig. 3.2.6, it is seen that the number of trapped particles increases as the vibration displacement amplitude at the tip increases. This is caused by the kinetic energy density increase near the sharp edges (lower end of the metal plates). A particle with a weight up to 256 milligrams can be sucked when the vibration displacement amplitude at the sharp edges is larger than 34 μm. Among all the particles, it is observed that it is the easiest to trap pill *C* which has a cubic shape. From Fig. 3.2.2 and Table 3.2.1, it is seen that pill *C* with cubic shape (1.1662 g/cm^3) has larger density than pill *E* with cylindrical shape (0.8541 g/cm^3) and similar density to pill *A* with spherical shape (1.1975 g/cm^3). This means that particle shape affects the trapping capability, and it is more difficult to suck spherical and cylindrical particles than cubic particles for a given vibration amplitude.

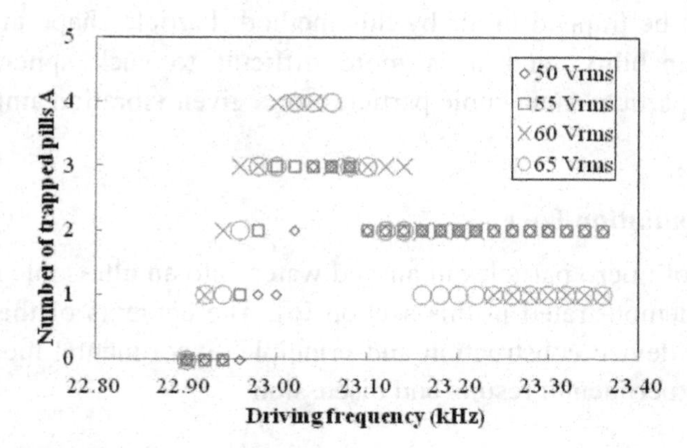

Fig. 3.2.5. Measured relationship between the number of trapped pills A and driving frequency in air under different input voltages.

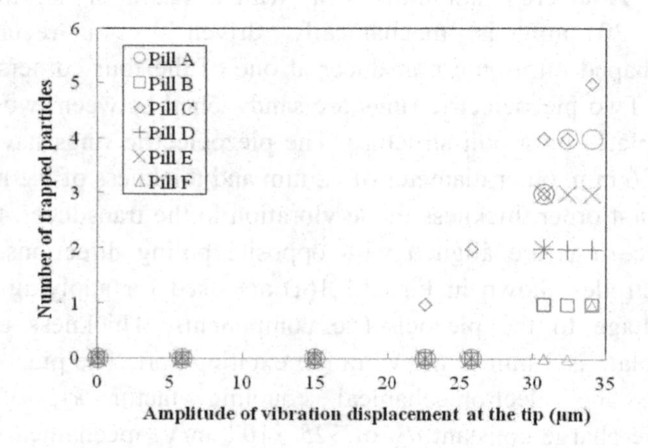

Fig. 3.2.6. Measured relationship between the number of trapped particles and the *y*-direction vibration displacement amplitude (0-peak) at the sharp edges for different particle types in air.

3.2.3 *Summary*

Acoustic radiation lines can be employed to suck and trap particles. Particles such as medicine pills with a weight up to 256 milligrams per particle can be trapped in air by this method. Particle shape affects the trapping capability, and it is more difficult to suck spherical and cylindrical particles than cubic particles for a given vibration amplitude.

3.3 On a Radiation Face

Collection of micro particles in air and water onto an ultrasonic radiation surface is demonstrated in this section [6]. The contents of this section include the device construction and principle, experimental method and particles, experimental results and discussion.

3.3.1 *Construction and principle*

Fig. 3.3.1 shows the structure and size of the actuator used for particle collection. A tapered aluminum strip with a width of 4.5 mm and length of 30 mm is mechanically driven by a rectangular-sandwich-shaped ultrasonic transducer at one of the four corners of the transducer. Two piezoelectric rings are sandwiched between two square aluminum plates by a bolt structure. The piezoelectric rings have inner diameter of 6 mm, outer diameter of 12 mm and thickness of 1.2 mm. To excite the first order thickness mode vibration in the transducer, the two piezoelectric rings are aligned with opposite poling directions. Two copper electrodes shown in Fig. 3.3.1(a) are used for applying an AC driving voltage to the piezoelectric components. Thickness of each aluminum plate is 3 mm in the vibration exciting part. The piezoelectric rings have the electromechanical coupling factor k_{33} of 0.71, piezoelectric charge constant d_{33} of 325×10^{-12} m/V, mechanical quality factor Q_m of 2000, and dissipation factor $tan\ \delta$ of 0.3. According to measurement, vibration displacement amplitude (0-peak) larger than 5 μm can be generated at the tip of the metal strip; the corners of the sandwich structure have a vibration stronger than the average along the edges of the square plates. The larger vibration at the corners is due to the smaller mechanical constraints on them.

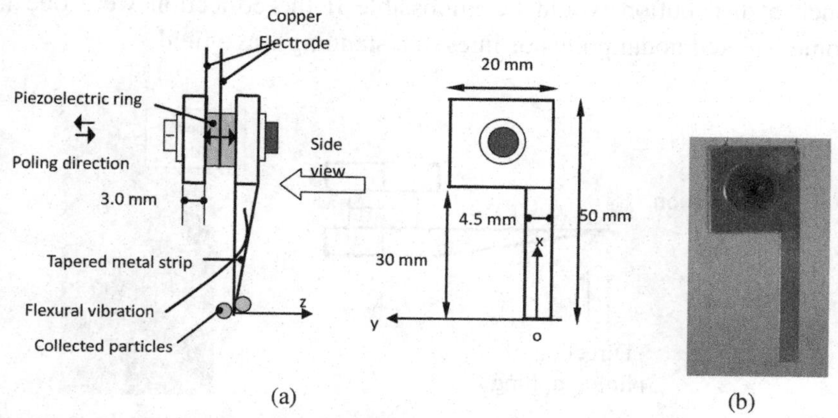

Fig. 3.3.1. Structure and size of the ultrasonic actuator with a tapered metal strip. (a) Schematic diagram; (b) photograph.

The first order thickness mode of the transducer was excited at approximately 105 kHz, and resulted in a flexural vibration of the aluminum strip. It was experimentally found that the vibration amplitude is the maximum at the tip ($x = 0$), and it decreases as x increases in the vicinity of the tip. This non-uniform vibration distribution and the relatively large vibration of the aluminum strip make it possible to obtain a large vibration at the tip. Due to the vibration, a sound field is formed in the fluid around the aluminum strip. Based on the analyses in Section 3 of Chapter 2, acoustic radiation force can be generated on particles near the vibrating strip, pushing the particles onto the vibrating strip surface.

To confirm the principle, an experiment was conducted, as shown in Fig. 3.3.2. In the experiment, the radiation surface was perpendicular to gravitation. Trapping of shrimp eggs in air on the lower radiation surface was observed. Fig. 3.3.2(b) is a photograph of trapped particles, taken from the direction shown in Fig. 3.3.2(a). Collection of particles in water also was experimentally investigated. Fig. 3.3.3 shows the collection of mint seeds in water by the metal strip in vibration. It was observed that the distribution of trapped particles is quite uniform in the trapping area. This provides experimental evidence to support the operating principle that the trapping force is caused by the near field on the radiation face.

Such a distribution would be impossible if the collection were due to some isolated nodal points or lines of a standing wave field.

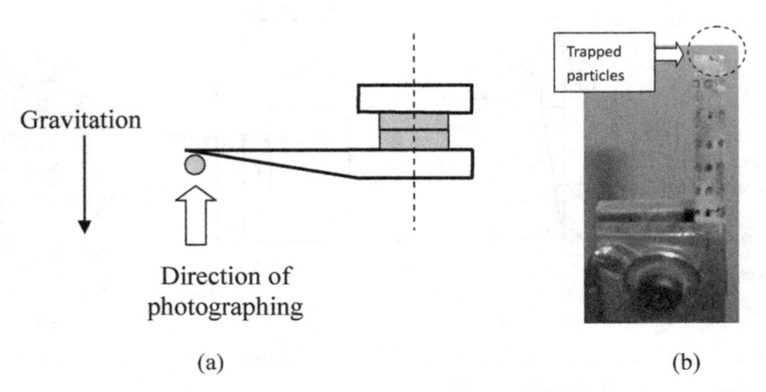

Fig. 3.3.2. (a) An experimental setup to confirm the existence of a trapping force perpendicular to an ultrasonic radiation face; (b) a photograph indicating the existence of a trapping force perpendicular to a radiation face.

Fig. 3.3.3. Particle trapping in water on the surface of a metal strip in vibration.

3.3.2 *Experimental method and particles*

The actuator was driven by a high-speed power amplifier (HSA4014, NF Corp., Yokohama, Japan), which received a sinusoidal signal from a function generator. Driving frequency of the actuator was around 105 kHz. Shrimp eggs, mint seeds and grass seeds were used in the experiments. Their density, volume and weight per particle are shown in

Table 3.3.1. In the table, the density was calculated from the measured volume and weight of a collection of particles; the volume per particle was calculated from the shape and size of a particle measured by microscope; the weight per particle was calculated from the density and volume per particle; the equivalent diameter was calculated from the volume per particle by assuming the particle to be an ideal sphere.

Table 3.3.1. Density, volume and weight per particle.

Materials	Density (g/cm^3)	Volume (10^{-3} cm^3)	Weight (g)	Equivalent diameter (μm)
Shrimp eggs	0.55	0.00671	3.7×10^{-6}	234
Mint seeds	0.70	0.0704	4.93×10^{-5}	513
Grass seeds	0.53	0.72	3.8×10^{-4}	1112

The capability of the actuator to collect these particles in air and water was investigated. In the experiments of collecting particles in air, the vibrating strip was moved onto the surface of a collection of particles. And in the experiments of collecting particles in water, the vibrating strip was submerged into water to approach particles. To count the collected particles, trapped particles were released into a box by switching off the input voltage.

During the experiments, it was difficult to measure the vibration at the tip directly. To investigate the effect of vibration of the actuator on the number of collected particles, the relationship between the motional current I_m of the actuator and vibration displacement amplitude w (0-peak) at the tip was measured for the actuator vibrating in air. The results are shown in Fig. 3.3.4. The motional current can be calculated from the measured input voltage, input current, phase difference between the input voltage and current, and clamped capacitance. As expected, the motional current is linearly proportional to the vibration displacement amplitude w when w is not too large. From Fig. 3.3.4, it is known that w/I_m is about 44.7 μm/A. This coefficient was used to calculate the vibration displacement amplitude w from the motional current.

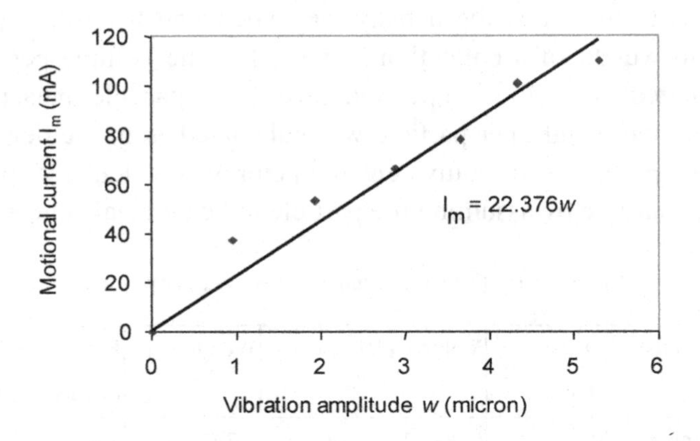

Fig. 3.3.4. Motional current of the actuator versus vibration displacement amplitude (0-p) of the tip.

3.3.3 *Experimental results and discussion*

Fig. 3.3.5 is a photograph showing the concentration of shrimp eggs on the tip of the metal strip in air. The length direction of the metal strip is parallel to gravitation.

Fig. 3.3.6 shows the relationship between the number of collected shrimp eggs and driving frequency at different driving voltages. The averaged number of collected particles of several experiments was used in the figure. It is seen that the number of collected particles reaches a maximum at some driving frequency for a given voltage. This driving frequency is the resonance frequency at the given voltage. It is also seen that when the driving voltage is less than 20 Vrms, this maximum number increases with increasing the driving voltage; while when the driving voltage is greater than 20 Vrms, this maximum number decreases with increasing the driving voltage. The former is due to the vibration increase, and the latter is due to the acoustic streaming which flows from the radiation surface to far field and flushes some of the collected particles away. The streaming was easily observed in water by using black ink film. However, it is not easy to clearly observe this streaming in air.

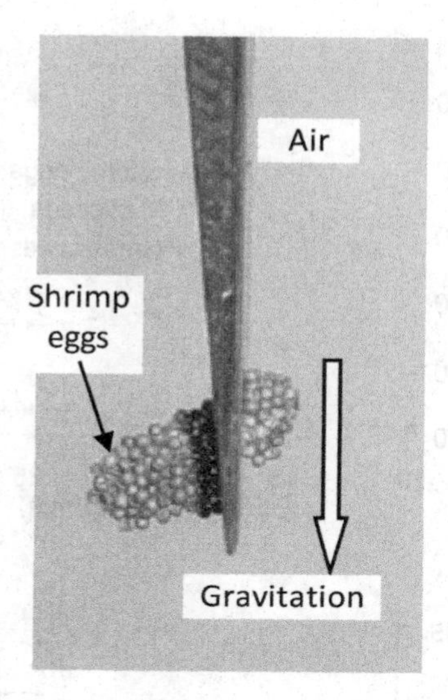

Fig. 3.3.5. Collection of shrimp eggs on the surface of the tip in air. The length direction of the metal strip is parallel to gravitation.

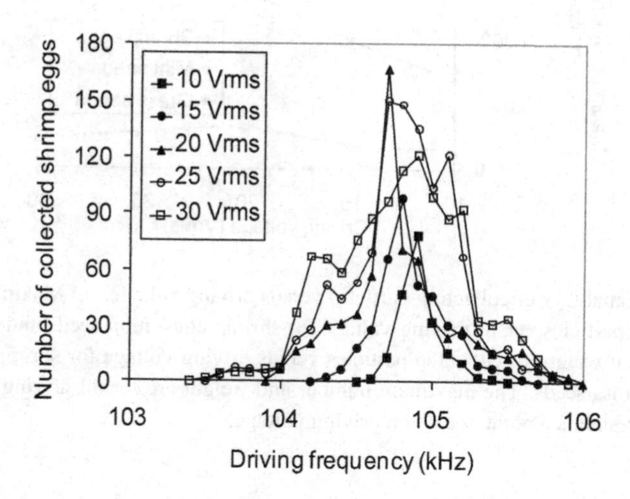

Fig. 3.3.6. Number of collected shrimp eggs versus driving frequency at different driving voltages.

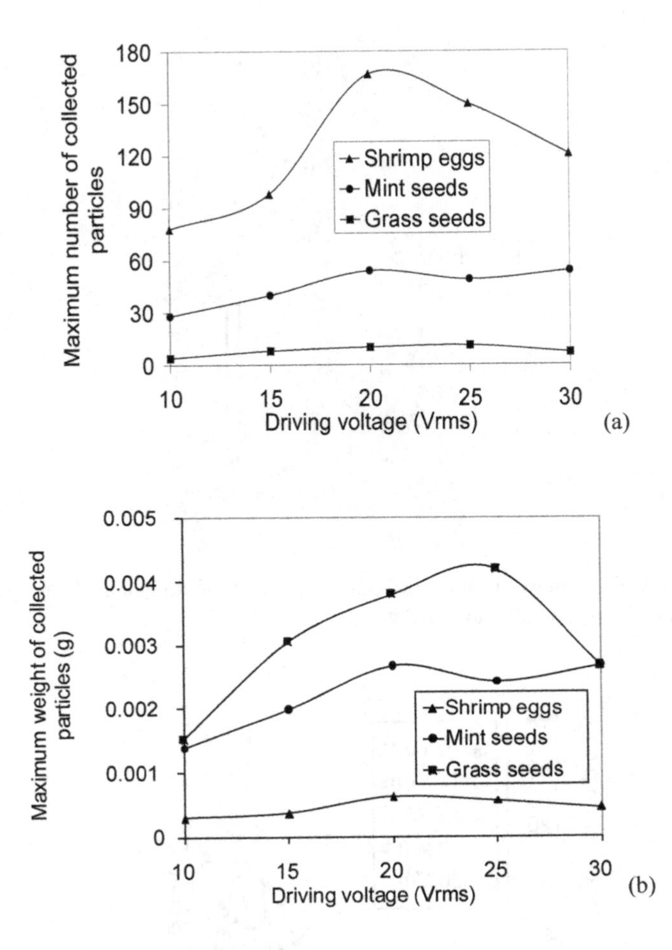

Fig. 3.3.7. Capability of collecting particles versus driving voltage. (a) Maximum number of collected particles versus driving voltage for shrimp eggs, mint seeds and grass seeds. (b) Maximum weight of collected particles versus driving voltage for shrimp eggs, mint seeds and grass seeds. The maximum number and weight are for all driving frequencies around the resonance point at a given driving voltage.

Similar results were obtained for mint seeds and grass seeds in air. Fig. 3.3.7(a) shows the dependence of the maximum number of collected particles on the driving voltage for shrimp eggs, mint seeds and grass seeds in air. It is seen that shrimp eggs are the easiest to collect among shrimp eggs, mint seeds and grass seeds; while grass seeds are the most difficult to collect. From Table 3.3.1, it is seen that among the three particles which have close densities, shrimp eggs have the smallest weight per particle while grass seeds have the largest. Therefore, for the particles with close densities, the smaller the weight per particle, the easier it is to collect. From Fig. 3.3.7(a), it is also seen that the driving voltage above which the maximum number of collected particles begins to decrease almost does not depend on particle type. This phenomenon can be explained as follows. The acoustic streaming which flows from the radiation face to far field cannot be generated unless the sound field is stronger than a critical one. Also, operating points indicated in Fig. 3.3.7 (a) correspond to the resonance states of the actuator thus driving voltage can solely determine the sound field around the radiation face. Therefore, when the driving voltage is less than a critical voltage corresponding to the critical sound field, the sound field is not large enough to generate a flow to affect the trapped particles; while when the driving voltage is larger than the critical voltage, the sound field starts to generate an acoustic streaming which can flush the trapped particles away. Furthermore, the critical sound field is independent of particle type. So the critical voltage is also independent of particle type as long as the number of collected particles is not large enough to affect the relationship between the sound field and driving voltage. Fig. 3.3.7(b) shows the dependence of the maximum weight of collected particles on the driving voltage for shrimp eggs, mint seeds and grass seeds. It is seen that collected grass seeds have the largest weight, while collected shrimp eggs have the smallest weight. The possible reason of this phenomenon is that grass seeds have the largest volume per particle and the trapping force per particle is the maximum.

For the large-density particles such as salt crystals (density ≈ 1.4 g/cm^3), few particles could be collected by the metal strip. This is because the trapping force on the particles is not large enough to generate a static frictional force to balance their weight.

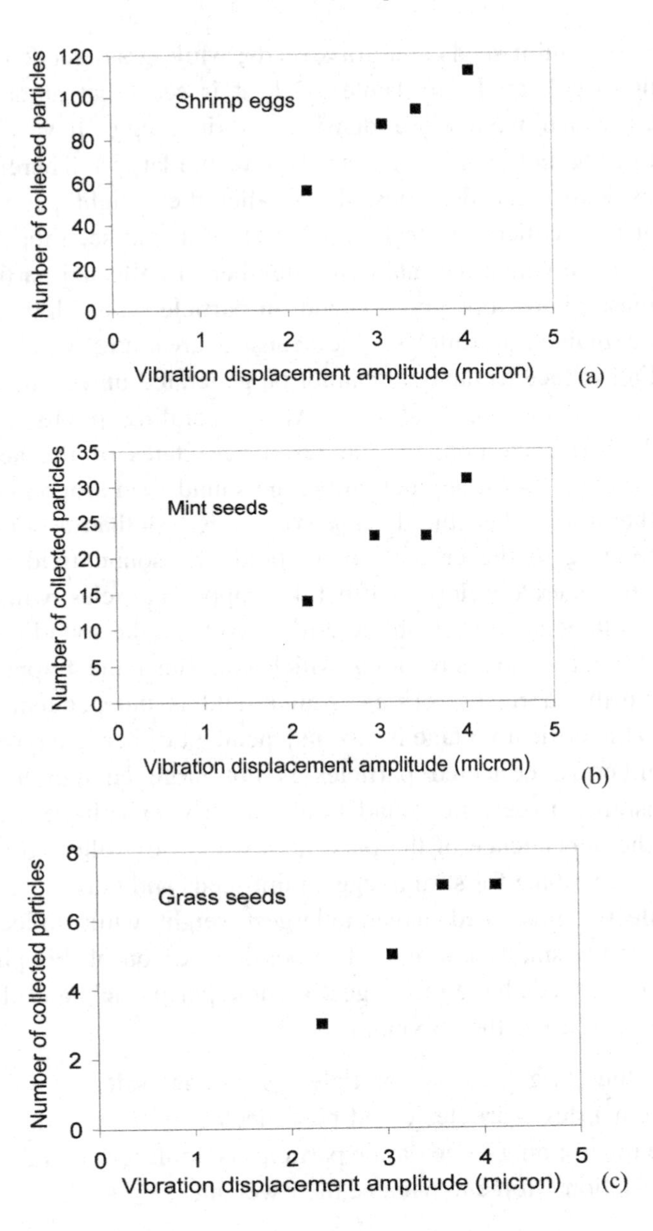

Fig. 3.3.8. Number of collected particles versus vibration displacement amplitude (0-p) of the tip. (a) Shrimp eggs; (b) mint seeds; (c) grass seeds. The length direction of the metal strip is parallel to gravitation.

(a) (b) (c)

Fig. 3.3.9. Collection of particles by the metal strip whose length direction is perpendicular to gravitation. (a) Shrimp eggs; (b) Mint seeds; (c) Grass seeds.

Figs. 3.3.8(a), (b) and (c) show the relationship between the number of collected particles and vibration displacement amplitude at the tip for shrimp eggs, mint seeds and grass seeds in air, respectively. It is seen that when the vibration displacement amplitude is not too large, the number of collected particles is approximately linearly proportional to the vibration displacement amplitude. Fig. 3.3.8 only shows the case at small vibration. When vibration of the tip is large, the number of collected particles decreases as the vibration increases because of the acoustic streaming from the radiation face.

In the above experiments, the length direction of the metal strip is parallel to gravitation. We define this operation as *P* mode. Tiny particles can also be stably trapped by the radiation face when the length direction of the metal strip is perpendicular to gravitation. We define this operation as *V* mode. Fig. 3.3.9 shows the collection of shrimp eggs, mint seeds and grass seeds by the actuator operating at *V* mode in air. The particles tend to be collected at the two corners of the tip. This is because acoustic radiation force pushing the particles onto the metal strip is larger near the two corners. Figs. 3.3.10(a), (b) and (c) show a comparison of the number of collected particles versus driving frequency between the operations at *V* mode and *P* mode. A driving voltage of 20 Vrms was used in the experiments. It is seen that the actuator operating at *V* mode has weaker capability to collect particles.

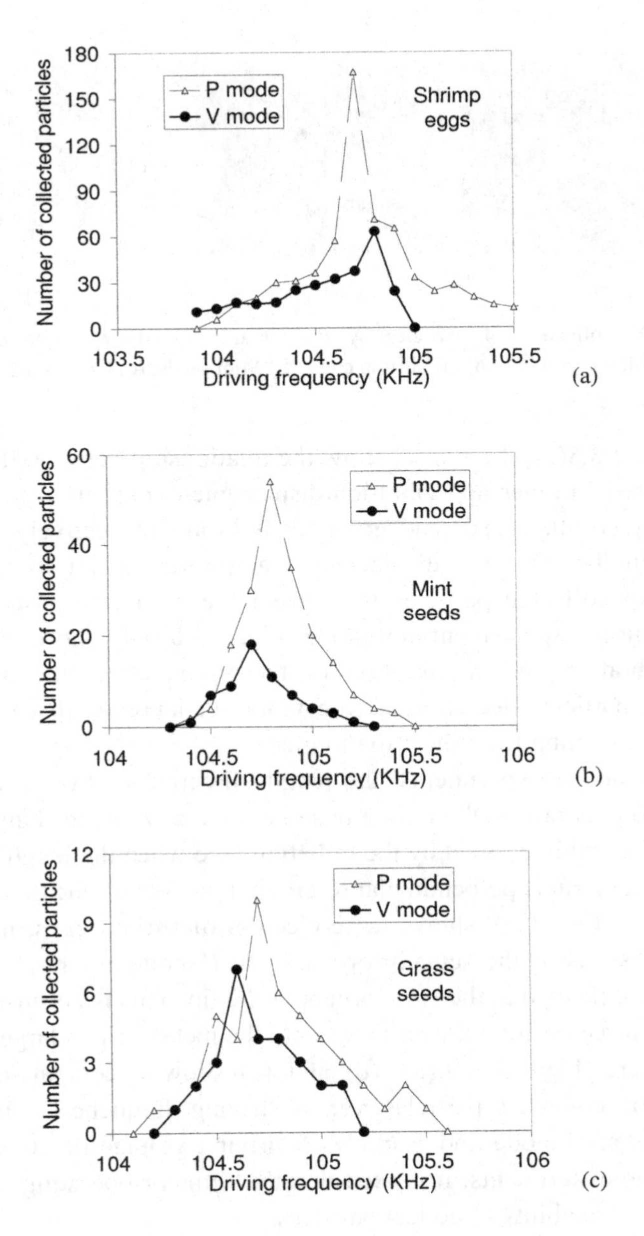

Fig. 3.3.10. Comparison of the number of collected particles versus driving frequency between the operation at *V* mode and *P* mode. (a) Shrimp eggs; (b) mint seeds; (c) grass seeds.

In the above experiments, particles to collect were in air. Particle collection in water was also experimentally investigated. Fig. 3.3.11 shows a two dimensional distribution of acoustic pressure near the radiation surface in water. The acoustic pressure was measured by a 1mm PVDF needle hydrophone (Precision Acoustics, UK), which made an angle of 45° with the measured surface. The outer diameter of the probe is about 1.5 mm. It is seen that the acoustic pressure decreases with increasing x near the tip, and is higher near the two sides than near the central region.

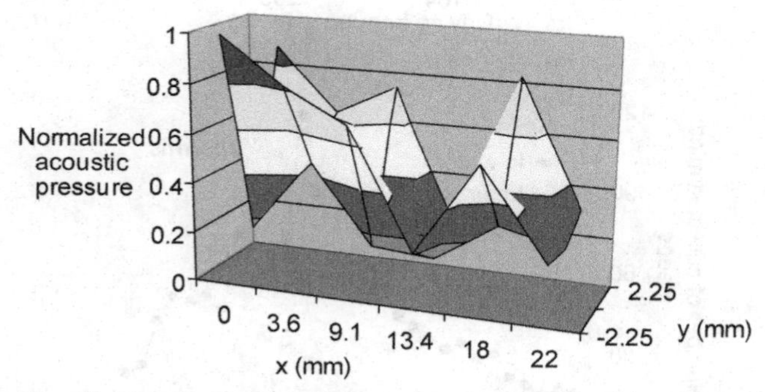

Fig. 3.3.11. Distribution of acoustic pressure near the metal strip.

Fig. 3.3.12 shows the frequency dependence of the number of mint seeds collected in water at different driving voltages, which was measured by the actuator operating at P mode. Mint seeds were collected from the bottom of a vessel with water. They were released into a box for counting when the depth of the metal strip submerged into water was about 15 mm. For comparison, frequency dependence of the number of collected mint seeds in air, which was measured by the actuator operating at P mode, is also shown in the figure. It is seen that more mint seeds can be collected in water than in air and the frequency range in which the actuator can collect particles is wider for the collection in water than in air. There are two possible reasons for these phenomena. One is that the matching of acoustic impedance between water and aluminum is better than that between air and aluminum. Another is that the buoyancy acting on the particles improves the capability of collecting particles.

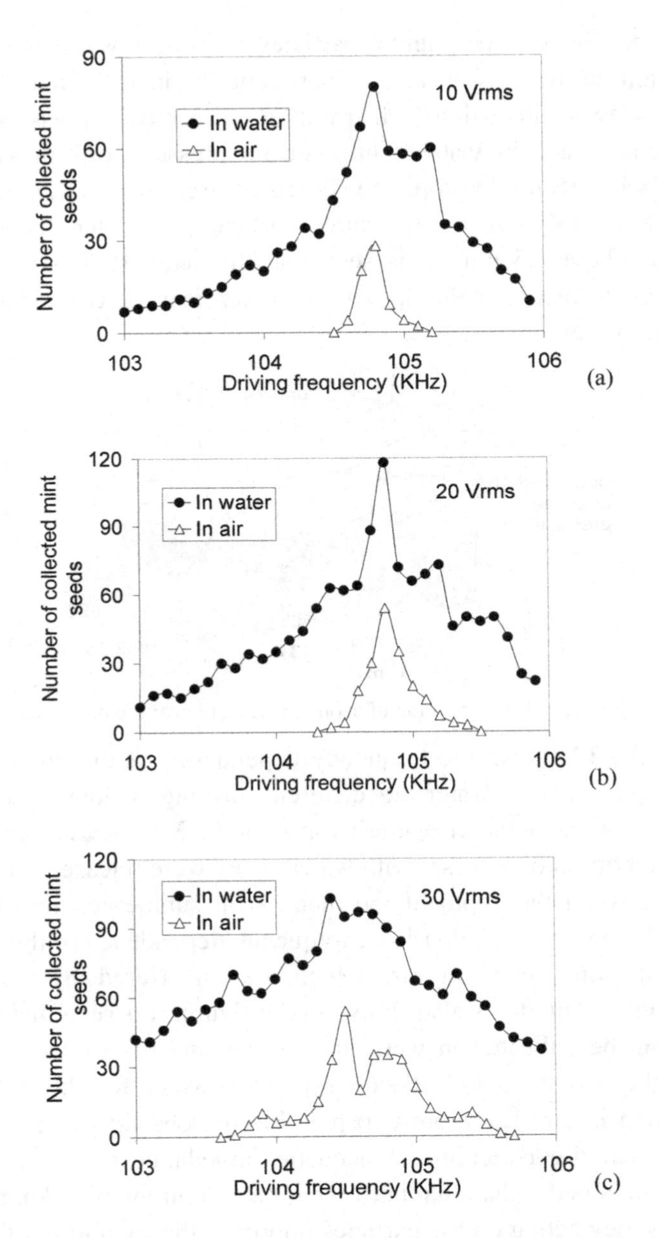

Fig. 3.3.12. Comparison of the number of collected mint seeds versus driving frequency in air and water at different driving voltages. The actuator operates at *P* mode.

3.3.4. *Summary*

The method of collecting small particles in air and water by an acoustic radiation face has been demonstrated. A sandwich-shaped ultrasonic transducer with a tapered metal strip is employed in the particle collection. Particles such as shrimp eggs, mint seeds and grass seeds may be collected and concentrated by the radiation surface of the metal strip when the sound field is appropriate. The capability of collecting particles depends on the weight per particle, vibration amplitude, and fluid around the particles. The capability increases as the vibration of the actuator increases, and is stronger in water than in air. However, this capability is limited by the acoustic streaming from the tip. When the vibration of the actuator is too large, the acoustic streaming may flush some of the trapped particles away.

3.4 π-shaped Ultrasonic Tweezers

The π-shaped ultrasonic tweezers which can trap tiny particles in air and water is demonstrated in this section. When it is used to trap particles in air, it operates in noncontact mode, that is, trapped particles are not in contact with the tweezers. When it is used to trap water droplet with particles inside, it operates in contact mode, that is, the water droplet is in contact with manipulating part of the tweezers. In this section, manipulation principles of the π-shaped ultrasonic tweezers are given

Table 3.4.1. Material constants of the piezoelectric rings used in the experiments.

k_t	$\tan\delta$ (%)	Q_m	d_{33} ($\times 10^{-12}$ m/V)	ρ (kg/m^3)	Y^e_{33} ($\times 10^{10}$N/m^2)	$\varepsilon^T_{33}/\varepsilon_0$	σ
0.48	0.3	2500	310	7800	6.6	1470	0.29

K_t is the electromechanical coupling factor of the thickness vibration mode, $\tan\delta$ is the dissipation factor, Q_m is the mechanical quality factor, d_{33} is the piezoelectric charge constant, ρ is the density, Y^E_{33} is the Young's modulus, $\varepsilon^T_{33}/\varepsilon_0$ is the relative permittivity, and σ is the Poisson's ratio.

after its structure is described, and then the trapping characteristics and their application examples are given.

3.4.1 *Construction and principle*

Figs. 3.4.1 and 3.4.2 show the structure of the π-shaped ultrasonic tweezers. Two piezoelectric rings with opposite poling directions are pressed onto two metal plates by a bolt structure, as shown in

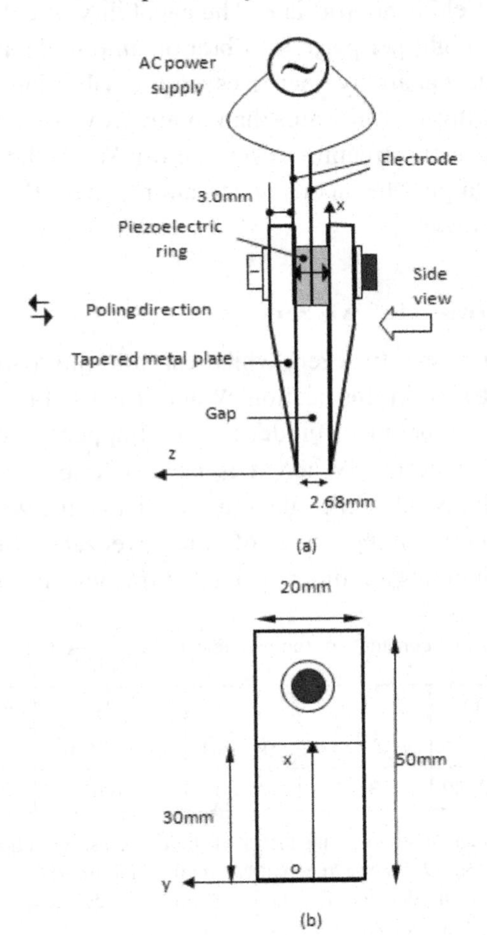

Fig. 3.4.1. Schematic diagram of the ultrasonic actuator with two tapered metal plates. (a) Structure; (b) side view.

Fig. 3.4.1(a). The piezoelectric rings have an outer diameter of 12 mm, inner diameter of 6 mm and thickness of 1.2 mm. The property constants of the piezoelectric rings are shown in Table 3.4.1. Between the two piezoelectric rings, and between one metal plate and its adjacent piezoelectric ring, there is a sheet of electrode, respectively. Due to the opposite poling directions, the 1st order thickness mode vibration can be excited if an AC voltage with a proper frequency is applied between the two sheets of electrode. The two metal plates made of aluminum are tapered in their lower parts, as shown in Fig. 3.4.1 (a) and (b). Each plate is 50 mm long and 20 mm wide. The thickness of the upper part of each metal plate is 3mm, and the length of the tapered part is 30mm. So the taper angle θ of the metal plates is about 5.7°. The operating frequency of the actuator is about 152.8 kHz.

The vibration velocity distribution of the metal plates was measured by laser scanning vibrometer when the upper part of the actuator was doing the 1st order thickness mode vibration, and the result is shown in Fig.3.4.3. In the figure, *Va* is the out-of-plane vibration velocity amplitude in the range of $0 \leq x \leq 30$ mm and -10 mm $\leq y \leq 10$ mm. It is seen that the two metal plates vibrate in a flexural mode with a gradient of amplitude in the *x* direction. The gradient of amplitude is due to the gradient of thickness of the metal plates in the *x* direction. It is also seen

Fig. 3.4.2. Photo of the actuator with two tapered metal plates.

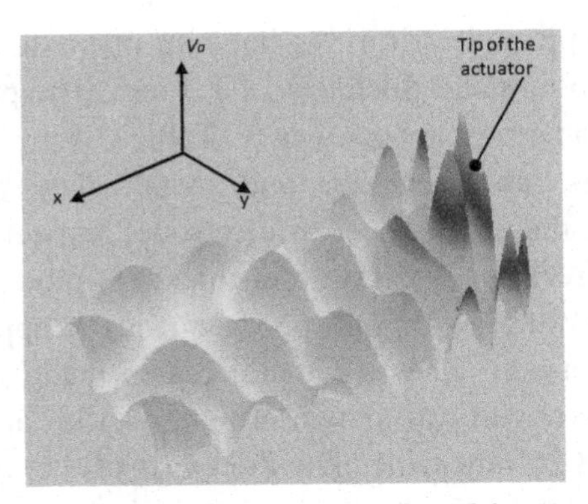

Fig. 3.4.3. Out-of-plane vibration amplitude distribution of the tapered actuator in air. *Va* is the out-of-plane vibration amplitude of the metal plates in air, and its maximum value in the figure is 8.0 mm/s. The measuring range is $0 \leq x \leq 30$ mm and -10 mm $\leq y \leq 10$ mm. All three coordinate axes are in linear scale.

that the wavelength increases as x increases. This is because the equivalent bending stiffness decreases as the plate thickness decreases.

Because of the flexural vibration of the metal plates, there is a standing wave sound field in the gap between the two plates. Also, sound field is strong in the lower part of the gap because vibration is quite strong in the lower part of the metal plates due to its small thickness. Here, the lower part refers to the part close to the plane of $x = 0$. Because of the acoustic radiation force generated by the sound field, small spheres in the gap can be trapped at the positions where the time-averaged force potential of the sound field is minimum. Because of the ultrasound induced weakening of intermolecular force in water, small droplets with particles inside can be extracted from water and transported in water by the tweezers. According to the analyses based on the force potential method (see Section 2.3 and Ref. 7), to trap a particle with the weight of *mg*, the maximum vibration velocity of the metal plate $V_{ib}(x)$ near to this particle must satisfy

$$V\,Y_c\,V_{ib}^2(x) \geq mg \tag{3.4.1}$$

where Y_c is a complicated function of the operating frequency, thickness of the gap, densities of the particles and fluid, and sound speeds in the particle and fluid.

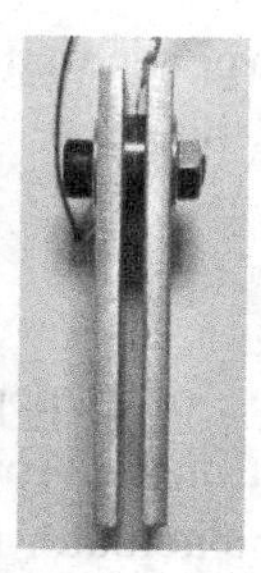

Fig 3.4.4. Photo of the actuator with two non-tapered metal plates.

To verify the contribution of the vibration gradient, we investigated the operation of the non-tapered actuator shown in Fig. 3.4.4, whose two metal plates have uniform thickness and other dimensions are the same

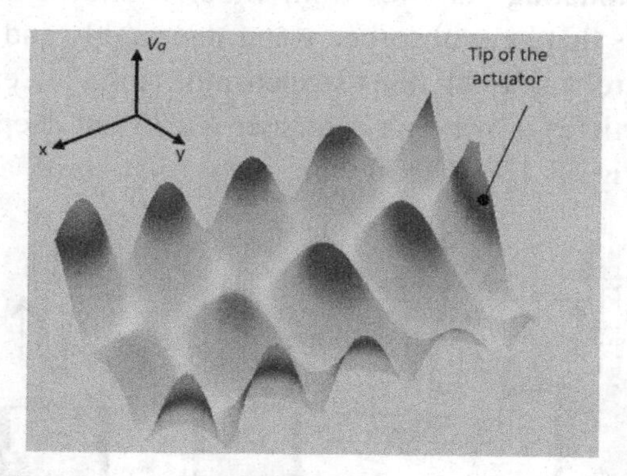

Fig. 3.4.5. Out-of-plane vibration amplitude distribution of the non-tapered actuator in air. V_a is the out-of-plane vibration amplitude of the metal plates in air, and its maximum value in the figure is 5.0 mm/s. The measuring range is 0≤x≤30mm and -10mm≤y≤10mm. All three coordinate axes are in linear scale.

as the ones of the tapered actuator. It was found that the particles used in our experiments could hardly be trapped or levitated in available input power range of the actuator. Furthermore, the vibration velocity distribution of this actuator was scanned, and the result is shown in Fig. 3.4.5. It is seen that the vibration distribution of the non-tapered actuator is evener along the x direction than that of the tapered actuator, and the vibration at the tip of the non-tapered actuator is smaller than that of the tapered one. For this reason, the vibration of the lower part of the non-tapered actuator is not large enough, and so is its standing wave sound field in the lower part of the gap.

3.4.2 *Experimental methods*

The experimental procedure of trapping particles is shown in Fig. 3.4.6(a). First, tapered part of the vibrating actuator is inserted into a collection of particles. Then, the actuator is lifted and moved onto a glass slider. Finally, the voltage applied to the actuator is switched off, and dropped particles on the glass slider are counted by microscope system (Olympus BX-51). Non-contact trapping of the particles is recorded by digital camera (Canon Power Shot S30). Similarly, the experimental procedure of extracting particles from water is shown in Fig. 3.4.6(b). The thyme seeds that have absorbed water thoroughly and have a density of 1.1g/cm^3 are used in the suspension. In the experiments of transporting particles in water, a container with 3 cm deep water is used, as shown in Fig. 3.4.7. The thyme seeds that have absorbed water

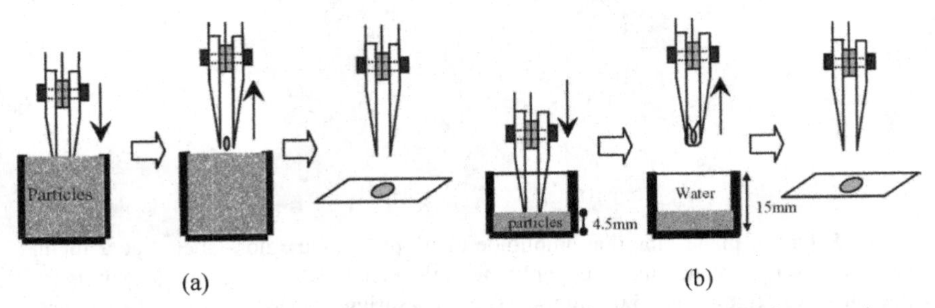

Fig. 3.4.6. Experimental procedures. (a) Noncontact trapping; (b) particle extraction.

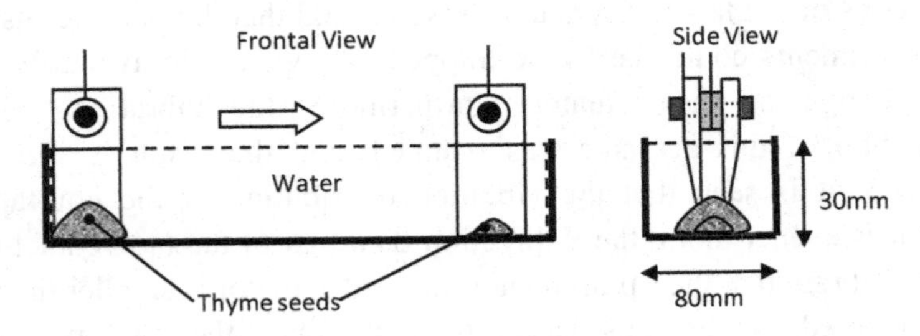

Fig. 3.4.7. Experimental method of the particle transportation in water.

thoroughly are linearly transported to a destination in water by the vibrating actuator. The thyme seeds that can reach the destination are collected and counted.

The tweezers is driven by a high-speed power amplifier (NF HSA4014), which receives a sine signal from a function generator. In the measurement of characteristics of the actuator, the input voltage and current, and phase difference between the input voltage and current are recorded. To get the vibration velocity from these data, the relationship between the motional current and the vibration velocity of the actuator is used, which can be obtained by measuring the vibration velocity, input voltage and current, and the phase difference between the input voltage and current. A laser scanning vibrometer system is used to measure the vibration mode pattern and vibration velocity of the actuator. The system consists of a scanning head (Polytec. OFV056), controller (Polytec. OFV 3001S), junction box (Polytec. PSV-Z-040-F) and PC with software PSV 7.11. In the following characteristics of the actuator, unless otherwise specified, the vibration velocity is the root mean square value at the origin of the Cartesian coordinate system shown in Fig. 3.4.1. According to measurement, the vibration velocity per unit motional current is about 9.2 m/s/A for the actuator operating in air. The particles used in the

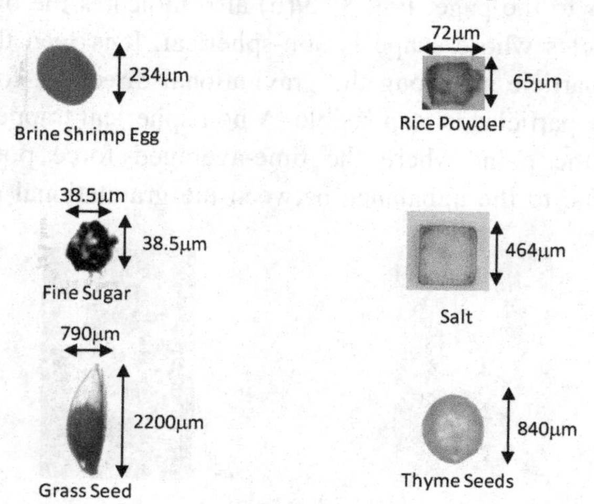

Fig. 3.4.8. Size and shape of the particles used in the experiments.

experiments are shown in Fig. 3.4.8. Their densities are shown in Table 3.4.2.

Table 3.4.2. Densities of the particles used in the experiments.

Materials	Density (g/cm^3)
Brine Shrimp Eggs	0.55
Fine sugar	0.53
Grass seeds	0.51
Rice powder	0.81
Fine salt	1.44
Thyme seeds	0.93

3.4.3 *Characteristics and discussion*

3.4.3.1 *Noncontact trapping*

It is experimentally confirmed that the tweezers can trap all of the particles listed in Table 3.4.2 or Fig. 3.4.8. Noncontact trapping of the brine shrimp eggs and grass seeds is shown in Fig. 3.4.9 (a) and (b), respectively. The black spot shown in Fig. 3.4.9(a) is the overlap of the images of several clusters of brine shrimp eggs along the direction perpendicular to the page. Fig. 3.4.9(b) also indicates the orientation of trapped particles whose shape is non-spherical. It is seen that the long axes of the particle are along the gravitational direction, for which the gravity of the particles is responsible. A non-spherical trapped particle is clamped at the point where the time-averaged force potential U is minimum. Due to the unbalance between the gravitational moments on

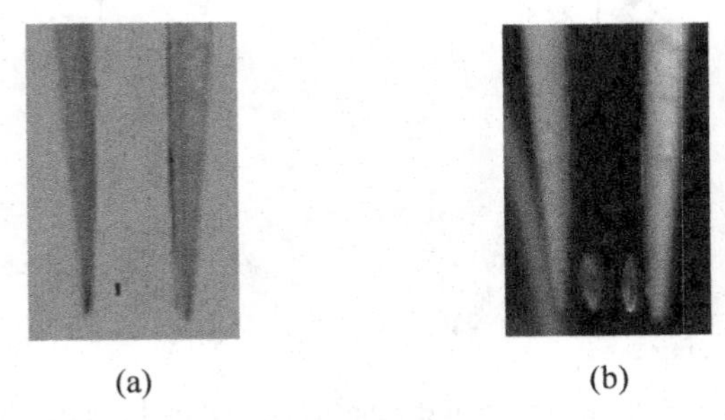

(a) (b)

Fig. 3.4.9. Photos of the noncontact trapping of particles in air. (a) Brine shrimp eggs; (b) grass seeds.

the two sides of the particle, which are separated by the clamping point, the levitated particle will rotate to the gesture at which the long axis of the particle is in parallel with the gravitational direction. At this gesture, the net gravitational moment on the particle is zero, and the particle is in equilibrium.

The relationships among the number of trapped particles, vibration velocity and input power were measured for brine shrimp eggs and grass seeds, and the result for brine shrimp eggs is shown in Fig. 3.4.10(a), and the one for grass seeds shown in Fig. 3.4.10(b). The number of trapped particles increases as the vibration velocity or input power increases. However, the number of trapped particles has a maximum value for the vibration velocity. When the vibration velocity or input power is higher than some value, the number of trapped particles decreases with increasing the input power. These phenomena may be explained as follows. To levitate a particle in the x direction, the local acoustic radiation force on the particle in the x direction must be larger than the weight of the particle; so the vibration velocity of the tweezers must be larger than a critical value. When the vibration velocity is low, only the lower part of the sound field in the gap can generate sufficient radiation force to levitate the particles in the x direction. As the vibration velocity increases, the vibration in the upper part of the sound field also increases, and this part can also generate large enough radiation force to levitate particles. So, the number of trapped particles increases as the vibration velocity increases. When the vibration velocity is too large, it can be observed that some of the levitated particles are ejected out of the actuator from the two sides of the upper part of the gap. The larger the vibration, the stronger the ejection is. So when the vibration is large enough, the number of trapped particles decreases as the vibration velocity increases. The acoustic streaming is responsible for this phenomenon. From Fig. 3.4.10 (a) and (b), it is also seen that the relationship between the number of trapped particles and the vibration velocity is a step function when the size and volume of trapped particles are relatively large, and the step change is not obvious when the size and volume of particles are relatively small. This is because the space available for trapping particles is limited in the tweezers.

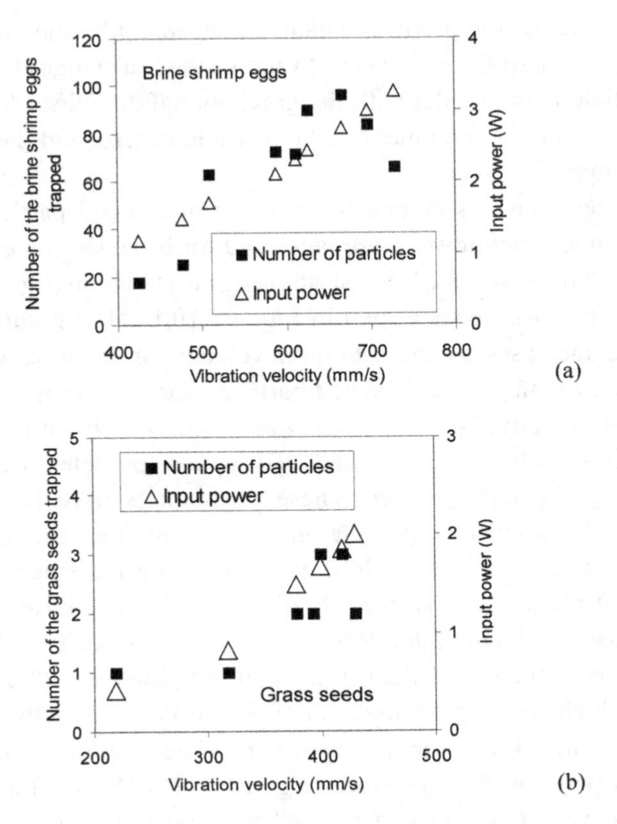

Fig. 3.4.10. Relationships among the number of the trapped particles, the vibration velocity and the input power in air. (a) Brine shrimp eggs; (b) grass seeds.

The dependence of the minimum vibration velocity to trap or levitate particles on the particle density was measured, and the result is shown in Fig. 3.4.11. It is seen that the minimum vibration velocity to trap particles increases as the particle density increases. This phenomenon can be explained by Eq. 3.4.1. From the equation, it is seen that to levitate a spherical particle with the weight of mg and volume of V, the vibration velocity of the metal plate $V_{ib}(x)$ must satisfy

$$V_{ib}(x) \geq \sqrt{\frac{mg}{VY_c}} = \sqrt{\frac{\rho_s g}{Y_c}} \tag{3.4.2}$$

So the minimum vibration velocity to trap particles increases as the particle density increases for spherical particles. Based on the derivation of Eq. 3.4.1, it is known that this conclusion does not hold when the particle shape is quite different from a sphere or the density ratio ρ_s (the density of particles)/ρ_0 (the density of acoustic medium) is close to one. According to our measurement, the minimum vibration velocity to trap the grass seeds and thyme seeds are 138 mm/s and 84 mm/s, respectively, and they do not satisfy the conclusion drawn from Fig.3.4.11. This is because the grass seeds have the shape very different from a sphere (see Fig. 3.4.8), and the thyme seeds have a relatively low density (see Table 3.4.2). From the above result, it is deduced that the tweezers may also be used in particle screening.

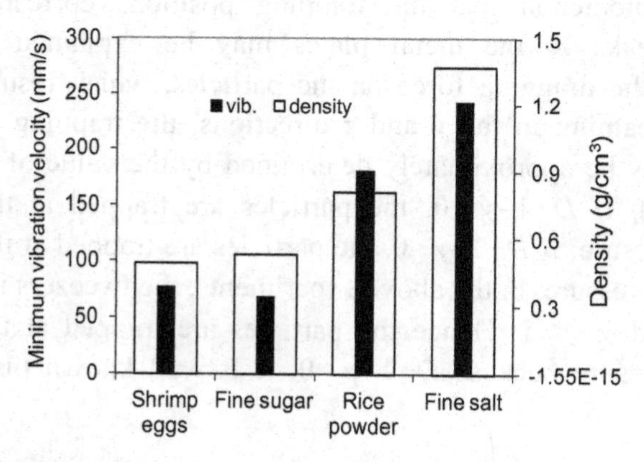

Fig. 3.4.11. The minimum vibration velocity to trap particles and the particles densities.

During our experiments, it was found that the trapping positions in the lower part of the gap in the *yz* plane were quite constant for different particles. Fig. 3.4.12 shows the distribution of trapped brine shrimp eggs, in which the trapping positions are shown by circle. Four positions P1, P2, P3 and P4 in the *yz* plane may trap the particles quite well, and P2 has the strongest trapping effect. By the comparison between Figs.3.4.3 and 3.4.12, it is known that these trapping positions correspond to the vibration peaks in the metal plates. P2 and P4 are due to the central vibration peaks of the metal plates at $y = 0$, as shown in Fig.3.4.3; while P1 and P3 are due to the side vibration peaks at $y = \pm 10$ mm. Because the

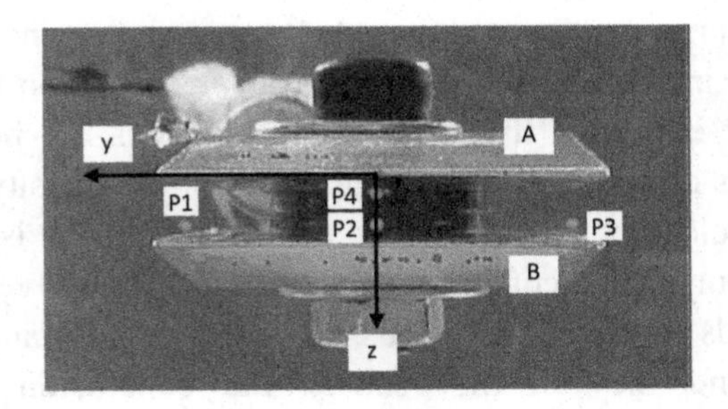

Fig. 3.4.12. Trapping positions in the *yz* plane near the tip of the actuator in air.

central vibration peaks are higher than the side ones, trapping effect at P2 is stronger than that at P1 and P3.

The phenomenon that the trapping positions correspond to the vibration peaks of the metal plates may be explained as follows. Neglecting the dragging force on the particles, which results from the acoustic streaming in the y and z directions, the trapping positions of particles may be approximately determined by the value of $D+1-\gamma$ (see Section 2.3). If $D+1-\gamma>0$, the particles are trapped at the nodes of acoustic pressure; if $D+1-\gamma<0$, the particles are trapped at the antinodes of acoustic pressure. In the above experiments, the tweezers is in air; thus $D\approx 1.5$ and $\gamma \ll 1$. Hence the particles are trapped at the nodes of acoustic pressure because $D+1-\gamma>0$. It is well known that there is a

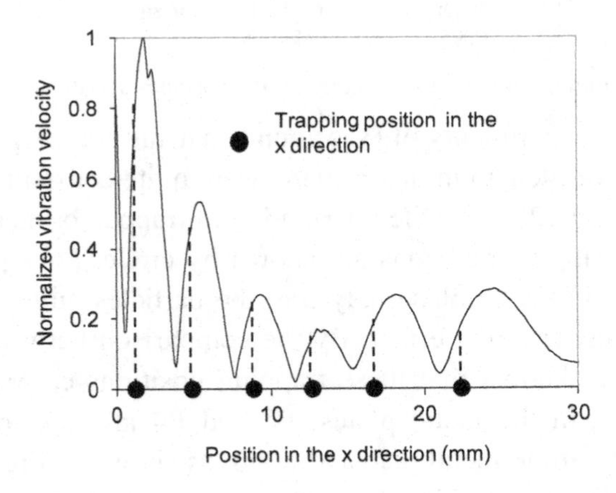

Fig. 3.4.13. Trapping positions and vibration distribution along the *x* direction in air.

spatial phase difference of $\pi/2$ between the acoustic pressure and vibration velocity of an acoustic field. So the particles are trapped at the antinodes of vibration velocity of the acoustic field. This is why the trapping positions correspond to the vibration peaks of the metal plates.

The measured trapping positions and vibration distribution of the metal plates along the x direction at $y = 0$ are shown in Fig. 3.4.13. Rice powder was used in the experiment. In the figure, circular black spots are used to indicate the trapping positions. It is seen that the particles are not exactly trapped at the positions where the corresponding vibration velocity of the metal plate is the maximum. The weight of the particles is responsible for this phenomenon. Due to the weight of the particles, the particles cannot be trapped at the nodal points of acoustic pressure because the acoustic radiation force in the x direction on the particles is zero at these nodal points. To get enough acoustic radiation force in the $+x$ direction to balance the weight, the particles must be trapped at the positions a little bit lower than the nodal points of acoustic pressure.

3.4.3.2 *Extraction and transportation*

Fig. 3.4.14 shows the dependence of the number of thyme seeds extracted from the suspension of thyme seeds and water on the input

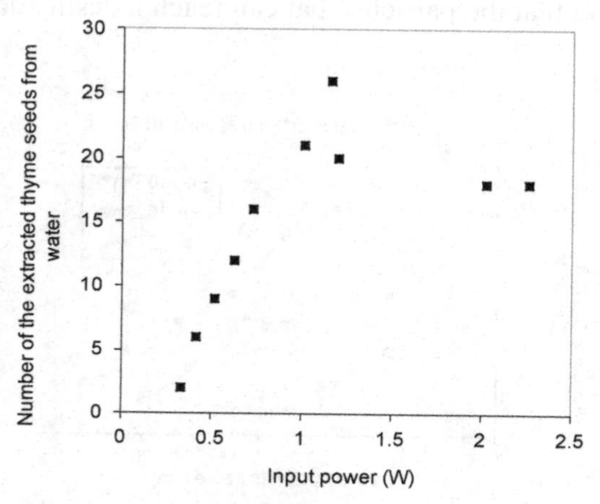

Fig. 3.4.14. Dependence of the number of extracted thyme seeds on the input power.

power of the tweezers in resonance. It is seen that the number of the extracted thyme seeds reaches a maximum value at the input power of 1.1W. The reason for the existence of this maximum value is similar with that for Fig. 3.4.10.

The number of thyme seeds, which can be linearly transported to a destination in water by the tweezers, depends on the transportation distance, and moving speed and input power of the tweezers. Fig. 3.4.15 shows the relationship among the number of thyme seeds that can reach a destination, the transportation distance and the input power when the tweezers in resonance moves at an average speed of 10.5 cm/s. It is seen that the number of thyme seeds that can reach a destination decreases as the transportation distance increases and the input power decreases. When the tweezers in resonance moving in water, there is relative motion between the water and tweezers in the gap, which may flush the particles away from the trapping positions where trapping effect is weak. The longer the transportation distance, the larger the particle loss during the transportation is. Also, as the input power decreases, trapping effect becomes weak and therefore the particle loss increases. Fig. 3.4.16 shows the relationship between the number of particles that can reach a destination in water and the moving speed of the actuator in resonance, for a constant input power (1090mW) and transportation distance (15cm). It shows that the particles that can reach a destination decease as

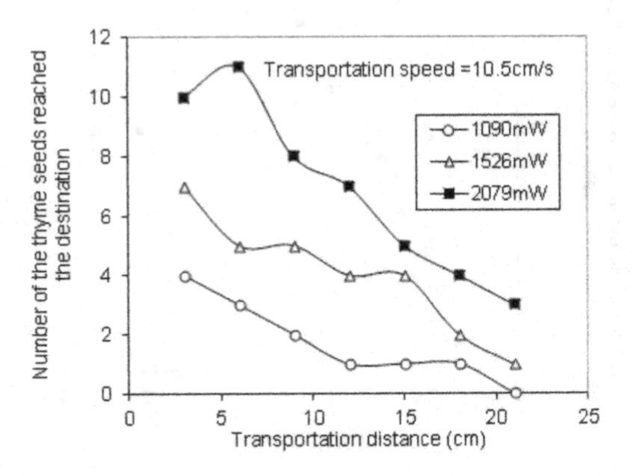

Fig. 3.4.15. Effects of the transportation distance and input power on the number of the thyme seeds that can reach a destination in water.

the moving speed of the tweezers increases. This is because as the moving speed increases, the dragging force on the particles increases.

3.4.4 *Summary*

The π-shaped ultrasonic tweezers can be used to trap, extract and transport tiny particles in air and water. In air, particles of spherical and non-spherical shapes with an average diameter from several tens μm to several hundred microns, such as shrimp eggs, grass seeds, fine salt particles, fine sugar particles, and thyme seeds can be trapped in the gap of the tweezers. The trapped particles are levitated in the gap and do not contact with the actuator. Increasing the vibration velocity can increase the number of trapped particles. But trapping capability of the tweezers is limited at large vibration. Particle density and shape affect the minimum vibration velocity for the trapping. Particles can be extracted from water and transported in water by the tweezers. Increasing the vibration velocity may increase the particle number extracted from water. Similar to the noncontact trapping, extracting capability of the tweezers is also limited at large vibration. Increasing the vibration velocity and decreasing the transportation speed can increase the number of the particles that can reach a destination in water.

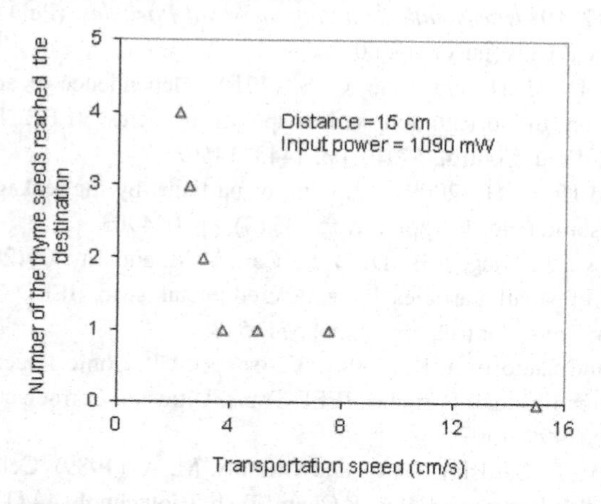

Fig. 3.4.16. Effect of the moving speed of the actuator on the number of the thyme seeds that can reach a destination in water.

3.5 Summary and Remarks

Micrometer scale or larger particles can be trapped by acoustic radiation force in contact or noncontact means. Based on the trapping, more manipulations such as extraction, transportation and separation may be realized. In addition to the methods listed described in this section, more methods can be used to trap micrometer scale or larger particles [8-13].

For the devices which use acoustic radiation force to trap particles, the acoustic streaming in the devices usually is a negative factor in manipulations because the streaming flushes trapped particles away. It is a big challenge to refrain or eliminate the acoustic streaming by optimizing the device structures and operation parameters such as the frequency of sound fields.

References

1. Hu, J. H., Tay, C., Cai, J. and Du, J. L. (2005). Controlled rotation of sound–trapped small particles by an acoustic needle, Appl. Phys. Lett., 87 (9), pp. 094104.
2. Hu, J. H., Tan, C. L. and Hu, W. Y. (2007). Ultrasonic microfluidic transportation based on a twisted bundle of thin metal wires, Sens. Actuators, A, 135 (2), pp. 811–817.
3. Liu, Y. Y. (2010) *Ultrasounic Trapping of Smalll Particles*, (PhD thesis, Nanyang Technological University), pp. 60.
4. Liu, Y. Y., Hu, J. H. and Zhao, C. S. (2010). Dependence of acoustic trapping capability on the orientation and shape of particles, IEEE Trans. Ultrason. Ferroelectr. Freq. Control, 57 (6), pp. 1443–1450.
5. Liu, Y. and Hu, J. H. (2009). Trapping of particles by the leakage of a standing wave ultrasonic field, J. Appl. Phys., 106 (3), pp. 034903.
6. Hu, J. H., Xu, J., Yang, J. B., Du, J. L., Cai , Y. M. and Tay, C. (2006). Ultrasonic collection of small particles by a tapered metal strip, IEEE Trans. Ultrason. Ferroelectr. Freq. Control, 53 (3), pp. 571–578.
7. Hu, J. H. and Santoso, A. K. (2004). A π–shaped Ultrasonic Tweezers Concept for Manipulation of Small Particles, IEEE Trans. Ultrason. Ferroelectr. Freq. Control, 51 (11), pp. 1499–1507.
8. Coakley, W. T., Bardsley, D. W. and Grundly, M. A. (1989). Cell manipulationin ultrasonic standing wave fields, J. Chem. Tech. Biotechnol., 44 (1), pp. 42–43.
9. Haake, A. and Dual, J. (2005). Contactless micromanipulation of small particles by an ultrasound field excited by a vibrating body, J. Acoust. Soc. Am., 117 (5), pp. 2752–2760.

10. Wu, J. (1991). Acoustic tweezers, J. Acoust. Soc. Amer., 89 (5), pp. 2140–2143.
11. Mitri, F. G. (2009). Langevin acoustic radiation force of a high–order Bessel beam on a rigid sphere, IEEE Trans. Ultrason. Ferroelectr. Freq. Control, 56 (5), pp. 1059–1064.
12. Yamakoshi, Y. and Noguchi, Y. (1998). Micro particle trapping by opposite phase ultrasonic traveling waves, Ultrasonics, 36 (8), pp. 873–878.
13. Chung, S. K., Rhee, K. and Cho, S. K. (2010). Bubble Actuation by Electrowetting–on–Dielectric (EWOD) and Its Applications: A Review, Int. J. Precis. Eng. Man., 11 (6), pp. 991–1006.

Chapter 4

Ultrasonic Extraction, Driving and Removal of Micro Solids

Extraction and spin driving of particles, rotary driving of small mechanical components, and dust removal for a solar panel by ultrasonic methods are demonstrated in this chapter.

4.1 Particle Extraction

This section presents a method for extraction of micro solid particles from a mixture of particles [1]. The contents of this section include the device construction, principle analyses, experimental materials and method, experimental and theoretical results, etc.

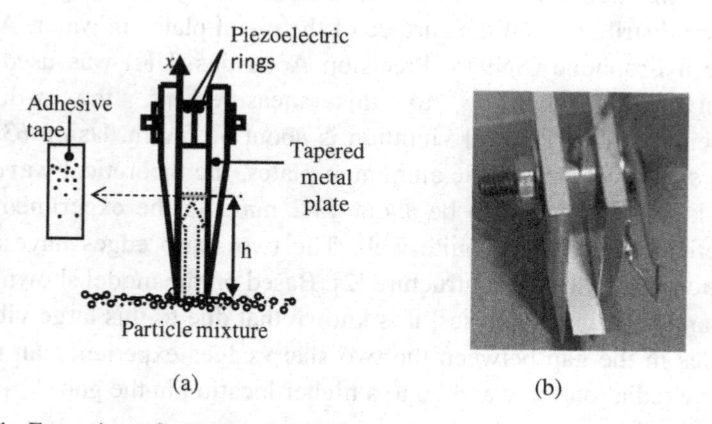

Fig. 4.1.1. Extraction of particles by the pumping effect in a π-shaped ultrasonic actuator. (a) Schematic diagram; (b) Photo.

4.1.1 *Construction*

Fig. 4.1.1 shows the method which exploits a π-shaped actuator to extract micro particles from a mixture of solid particles. Two metal plates made of aluminum sandwich a multilayer piezoelectric vibrator by a bolt structure. The multilayer piezoelectric vibrator is formed by two oppositely poled piezoelectric ceramic rings, which have an outer diameter of 12 mm, inner diameter of 6 mm and thickness of 2.4 mm. The piezoelectric rings have the electromechanical coupling factor k_{33} of 0.71, piezoelectric charge constant d_{33} of 325 $\times 10^{-12}$ m/V, mechanical quality factor Q_m of 2000, and dissipation factor $\tan\delta$ of 0.3. Each metal plate is 50 mm long and 20 mm wide. The thickness of the upper part of each metal plate is 3 mm, and the length of the tapered part is 30 mm. The operating frequency of the actuator is about 86 kHz. Small particles near the sharp edges of the metal plates can be pumped up to a higher location in the gap between the two metal plates. This phenomenon is termed pumping effect.

4.1.2 *Operating principle*

When the thickness vibration is excited in the upper part of the actuator, the lower part of the metal plates may operate at a flexural vibration mode. This flexural vibration was confirmed by scanning the sound pressure distribution on the surface of the metal plates in water. A 1 mm needle hydrophone (SN945, Precision Acoustics, UK) was used in the measurement. According to this measurement, the *x*-direction wavelength of the flexural vibration is about 66.6 mm. Using 6300 m/s as the sound velocity in the aluminum plates, the theoretical wavelength at 86 kHz is estimated to be about 73.2 mm. So the experimental and theoretical values agree quite well. The two sharp edges have a large vibration due to the taper structure [2]. Based on the model shown in Fig. 4.1.2 and following analyses, it is known that due to this large vibration, particles in the gap between the two sharp edges experience an upward acoustic radiation force and go to a higher location in the gap.

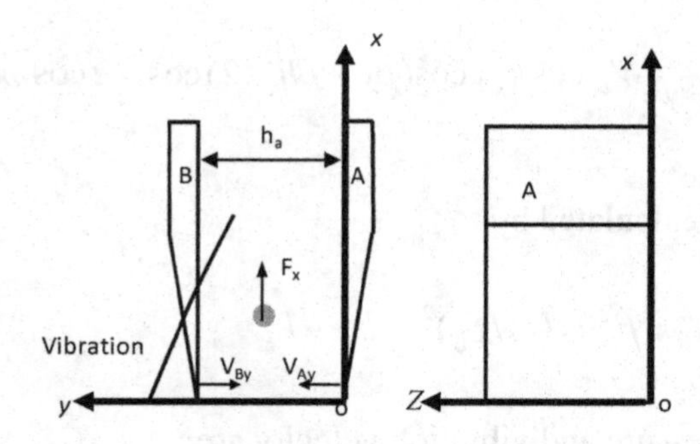

Fig. 4.1.2. A model for analyzing the operation.

4.1.2.1 *Analyses of acoustic radiation force*

The three dimensional sound field in the gap can be solved by the following wave equation and boundary conditions:

$$\frac{\partial^2 \varphi}{\partial t^2} = c_0^2 \nabla^2 \varphi \tag{4.1.1}$$

$$-\frac{\partial \varphi}{\partial y}\bigg|_{y=0} = V_{Ay} \tag{4.1.2}$$

$$-\frac{\partial \varphi}{\partial y}\bigg|_{y=h_a} = -V_{By} \tag{4.1.3}$$

where φ is the velocity potential in the gap, c_0 is the sound speed in the fluid, V_{Ay} and V_{By} are the vibration velocities of the two metal plates, respectively, and h_a is the thickness of the gap. Assuming

$$V_{Ay} = V_{By} = V_m \cos k_x x \cos k_z z \cos \omega t \tag{4.1.4}$$

where ω is the driving angular frequency, and k_x and k_z are the wave number along the x and z directions of the plate vibration, respectively. Neglecting the change of V_m near the sharp edges, approximately we have

$$\varphi = -\frac{1}{\beta \sin \beta h_a / 2} V_m \cos k_x x \cos(\beta y - \beta h_a / 2) \cos k_z z \cos \omega t \tag{4.1.5}$$

where β can be calculated by

$$\beta^2 = (\omega / c_0)^2 - k_x^2 - k_z^2 \tag{4.1.6}$$

The acoustic pressure and vibration velocity are

$$p = \rho_0 \frac{\partial \varphi}{\partial t} \tag{4.1.7}$$

$$\mathbf{V_f} = V_{fx}\mathbf{i} + V_{fy}\mathbf{j} + V_{fz}\mathbf{k} = -\nabla \varphi \tag{4.1.8}$$

where ρ_0 is the density of the fluid. Substituting (4.1.5) into (4.1.7) and (4.1.8), we obtain the acoustic pressure and vibration velocity in the gap:

$$p = \frac{V_m \rho_0 \omega}{\beta \sin \beta h_a / 2} \cos k_x x \cos(\beta y - \beta h_a / 2) \cos k_z z \sin \omega t \tag{4.1.9}$$

$$V_{fx} = \frac{-V_m k_x}{\beta \sin \beta h_a / 2} \sin k_x x \cos(\beta y - \beta h_a / 2) \cos k_z z \cos \omega t \tag{4.1.10}$$

$$V_{fy} = \frac{-V_m}{\sin \beta h_a / 2} \cos k_x x \sin(\beta y - \beta h_a / 2) \cos k_z z \cos \omega t \tag{4.1.11}$$

$$V_{fz} = \frac{-V_m k_z}{\beta \sin \beta h_a / 2} \cos k_x x \cos(\beta y - \beta h_a / 2) \sin k_z z \cos \omega t \tag{4.1.12}$$

Acoustic radiation force acting on a particle in the sound field is

$$\mathbf{F} = -\nabla U \tag{4.1.13}$$

where U is the time-averaged force potential of the sound field. When the wave number k (= ω/c_0) and the particle radius R satisfy $kR \le 1$, the force potential U is [3]

$$U = 2\pi R^3 \left[\frac{f_1}{3\rho_0 c_0^2} < p^2 > - \frac{f_2 \rho_0}{2} < V_f^2 > \right] \quad (4.1.14)$$

where < > means the time-averaged value, and f_1 and f_2 are

$$f_1 = 1 - \frac{\rho_0 c_0^2}{\rho_s c_s^2} \quad (4.1.15)$$

$$f_2 = \frac{2(\rho_s - \rho_0)}{2\rho_s + \rho_0} \quad (4.1.16)$$

where ρ_s and c_s are the density of and sound speed in the sphere, respectively.

In experiments, $\rho_s >> \rho_0$ because the fluid is air. Hence, $f_1 \approx 1$ and $f_2 \approx 1$. Also, according to measurement, the metal plates have no nodes and antinodes of vibration along the z direction for the actuator used in the experiments. Hence, $k_z = 0$. From these conditions and Eqs. (4.1.9)-(4.1.14), the acoustic radiation force F_x experienced by a particle in the x-direction at $y = h_a/2$ is

$$F_x = \frac{\pi R^3 \rho_0 V_m^2 k_x}{3 \sin^2 \beta h_a / 2} \sin 2k_x x = F_m \sin 2k_x x \quad (4.1.17)$$

Using Eq. (4.1.17) and measured k_x (= 95.3 rad/s), the distribution of F_x along the x direction is plotted, as shown in Fig. 4.1.3. It is seen that the acoustic radiation force F_x is positive in the range $0 < x < 16.5$ mm. This means that there is an upward acoustic radiation force acting on the particle in this range. Due to this force, the particles in the gap between the sharp edges may go to a higher location.

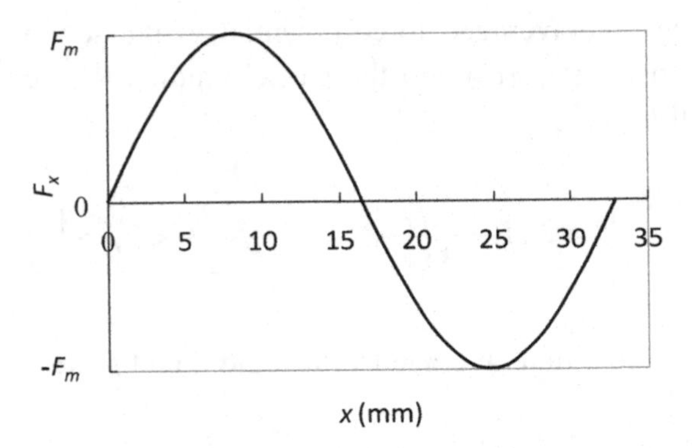

Fig. 4.1.3. The *x*-direction distribution of the x-component of acoustic radiation force acting on a particle in the middle of the gap.

4.1.2.2 *Analyses of the upward motion*

According to the energy conservation law, velocity of particles V at position x satisfies the following equation.

$$0.5mV_t^2 + E_a = 6\pi\eta R \int_{x'=0}^{x'=x} V dx' + mgx + 0.5mV^2 \tag{4.1.18}$$

where V_t is the average initial velocity of the particles, E_a the average energy transferred from the sound field to one particle, η the shear viscosity of the fluid, m the average mass per particle and g the gravitational acceleration, which is 9.8 m/s^2 under terrestrial gravity condition. Here, the assumption of creeping motion is adopted because the particle speed is not very large. So the drag experienced by each particle is $6\pi\eta RV$. To simplify the analyses, we let

$$E_0 = 0.5mV_0^2 = 0.5mV_t^2 + E_a \tag{4.1.19}$$

where E_0 and V_0 are the equivalent initial kinetic energy and initial velocity of a particle, respectively. From Eqs. (4.1.17) and (4.1.19), it is known that V_0 increases with increasing the actuator vibration. Substituting Eq. (4.1.19) into Eq. (4.1.18),

$$0.5mV_0^2 = 6\pi\eta R\int_{x'=0}^{x'=x} Vdx + mgx + 0.5mV^2 \qquad (4.1.20)$$

From Eq. (4.1.20),

$$x + m[\frac{V - V_0}{6\pi\eta R} - \frac{mg}{(6\pi\eta R)^2}\ln\frac{6\pi\eta RV + mg}{6\pi\eta RV_0 + mg}] = 0 \qquad (4.1.21)$$

At the maximum height h_{max} which the pumped up particles can reach, $V = 0$. Therefore, h_{max} can be calculated by the following equation:

$$h_{\max} = \frac{mV_0}{6\pi\eta R} - \frac{m^2 g}{(6\pi\eta R)^2}\ln(1 + \frac{6\pi\eta RV_0}{mg}) \qquad (4.1.22)$$

Using this equation, h_{max} versus m was calculated for three different radii at $E_0 = 2\times10^{-9}$ J, and the results are shown in Fig. 4.1.4. E_0 is chosen to be 2×10^{-9} J in order to explain the experimental phenomenon recorded in Table I (see section 4.1.3). η of air is 0.000018 Pa·s in the calculation. It

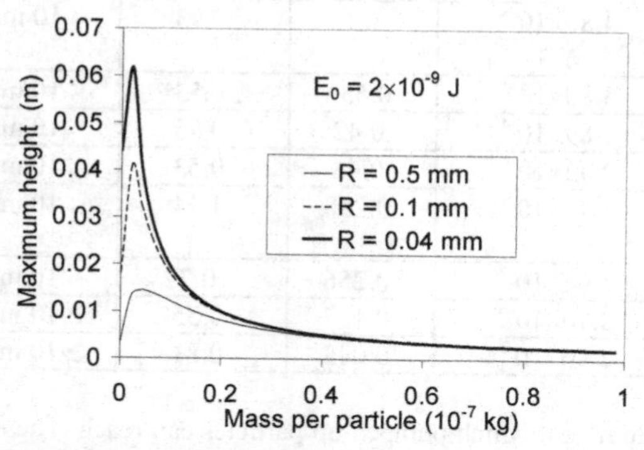

Fig.4.1.4. Effect of the mass per particle on the maximum height which a particle can reach at different particle radii and a given initial kinetic energy.

is seen that particles with smaller mass and radius can reach a higher location. For this reason, particles with smaller mass and radius can be extracted from a particle mixture.

4.1.3 *Experimental materials and method*

To investigate the pumping effect, the particles shown in Table 4.1.1 were used. The table also lists the mass, radius and density of the particles. The particle density was obtained by measuring the total volume and mass of a collection of particles. The mass per particle was calculated from the particle density and the volume which was calculated from the particle size which was obtained by a microscope system (Olympus BX-51, Olympus). In the experiments of extraction, two particle mixtures were used. One is a uniform mixture of shrimp eggs and fine salt crystals with a volume ratio of 1:1. Another is a uniform mixture of shrimp eggs and grass seeds with a volume ratio of 0.5:2.5. Adhesive tape was inserted into the gap between the two metal plates to collect the pumped up particles, as shown in Fig. 4.1.1.

Table 4.1.1. Pumping effect in the actuator for different particles.

	Mass per particle m (g)	Radius R (mm)	Density (g/cm^3)	Pumping effect h_{max}	
				Exp.	Cal.
Flying Colors seeds	4.89×10^{-4}	0.5	0.93	< 10 mm	0.4 mm
Grass seeds	3.80×10^{-4}	0.556	0.53	< 10 mm	0.5 mm
Thyme seeds	2.89×10^{-4}	0.42	0.93	< 10 mm	0.7 mm
Celery seeds	2.43×10^{-4}	0.48	0.53	< 10 mm	0.8 mm
Fine salt crystals	1.44×10^{-4}	0.288	1.44	< 10 mm	1.4 mm
Mint seeds	4.94×10^{-5}	0.256	0.7	< 10 mm	4.0 mm
Shrimp eggs	3.70×10^{-6}	0.117	0.55	≥ 10 mm	32.2 mm
Rice powder	2.46×10^{-7}	0.042	0.81	≥ 10 mm	60.2 mm

h_{max} is the maximum height which pumped up particles can reach. The experimental h_{max} was obtained within the allowed power range of the actuator. The theoretical h_{max} was obtained with the assumption that the initial kinetic energy of a particle is 2×10^{-9} J.

The actuator was driven by a high-speed power amplifier, which received a sinusoidal voltage from a function generator. The phase difference between the input voltage and current of the actuator was kept zero. Under this experimental condition, the input electric power of the actuator is approximately proportional to the mechanical output power, which is proportional to the square of vibration velocity at the sharp edges. Time period of pumping particles was about 3 seconds in each measurement. Also, the number of extracted particles shown in the following figures is the average value of several measurements.

4.1.4 *Results and discussion*

To verify the operating principle described in Section 4.1.2, the pumping effect was investigated by different particles, and the results are shown in Table 4.1.1. It can be seen that the particles with smaller mass and radius such as shrimp eggs and rice powder can be effectively pumped up to a height larger than 1 cm, and the particles with mass per particle larger than 5×10^{-5} gf cannot be pumped up to a height larger than 1 cm within allowed power range of the actuator. This result qualitatively verified the principle proposed. h_{max} was calculated for those particles at $E_0 = 2\times10^{-9}$ J, and the results are also shown in Table 4.1.1. It is seen that the theoretical results can well explain the experimental phenomena.

The dependence of the number of extracted shrimp eggs on the input electric power and on adhesive tape height h was experimentally investigated for a mixture of shrimp eggs and fine salt crystals and mixture of shrimp eggs and grass seeds. The results are shown in Fig. 4.1.5. In the experiment, there were no fine salt crystals and grass seeds collected by the adhesive tape when the adhesive tape height $h \geq 1$ cm. The input power was increased by increasing the input voltage and keeping the actuator at resonance; hence increasing the input power increases the vibration. It is seen that increasing the actuator's vibration and decreasing the adhesive tape height all increase the number of extracted shrimp eggs. As the vibration increases, the kinetic energy transferred to the particles around $x = 0$ and the work done by the sound field on the particles during the rise all increase. Hence, more shrimp eggs can reach the adhesive tape as the vibration increases.

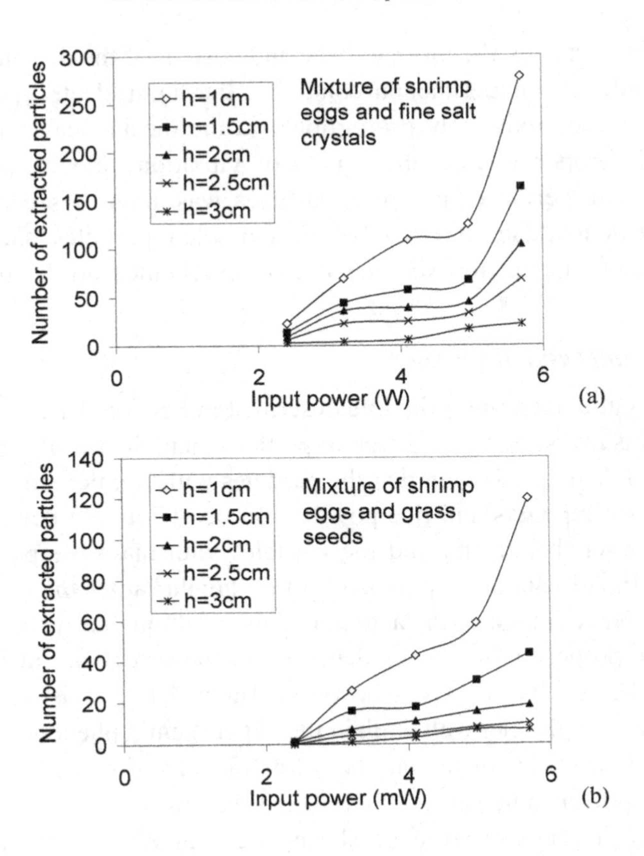

Fig. 4.1.5. Number of extracted shrimp eggs versus input electric power of the actuator for different adhesive tape height. (a) Extraction from a mixture of shrimp eggs and fine salt crystals; (b) Extraction from a mixture of shrimp eggs and grass seeds. *h* represents the adhesive tape height.

Figs. 4.1.6(a)-(d) show the effect of other particles in the mixture on the extraction of shrimp eggs at different input power. The input power was increased by increasing the input voltage and keeping the actuator at resonance; hence increasing the input power increases the vibration. In the experiment, there were no fine salt crystals and grass seeds collected by the adhesive tape because the adhesive tape height $h \geq 1$cm. For comparison, the number of shrimp eggs extracted from a collection of pure shrimp eggs versus the adhesive tape height is also given in the figures. From the figures, it is seen that fine salt crystals in the mixture enhance the extraction capability, and grass seeds weaken it. Exact mechanism for this phenomenon is still unknown.

h_{max} versus the initial x-direction velocity V_0 of a pumped up shrimp egg was calculated in air, water, steam and oxygen, and the results are shown in Fig. 4.1.7. In the calculation, the viscosity of air, water and steam is 0.000018 Pa·s, 0.001 Pa·s, 0.000013 Pa·s and 0.0002 Pa·s, respectively. It shows that the maximum height which a shrimp egg can reach decreases with increasing the viscosity, and it is difficult to pump up shrimp eggs in water due to a large viscosity of water. Experiments were conducted to observe the behavior of shrimp eggs in water suspension in the gap. It was observed that the shrimp eggs suspended in water could not be pumped up within allowed power range of the

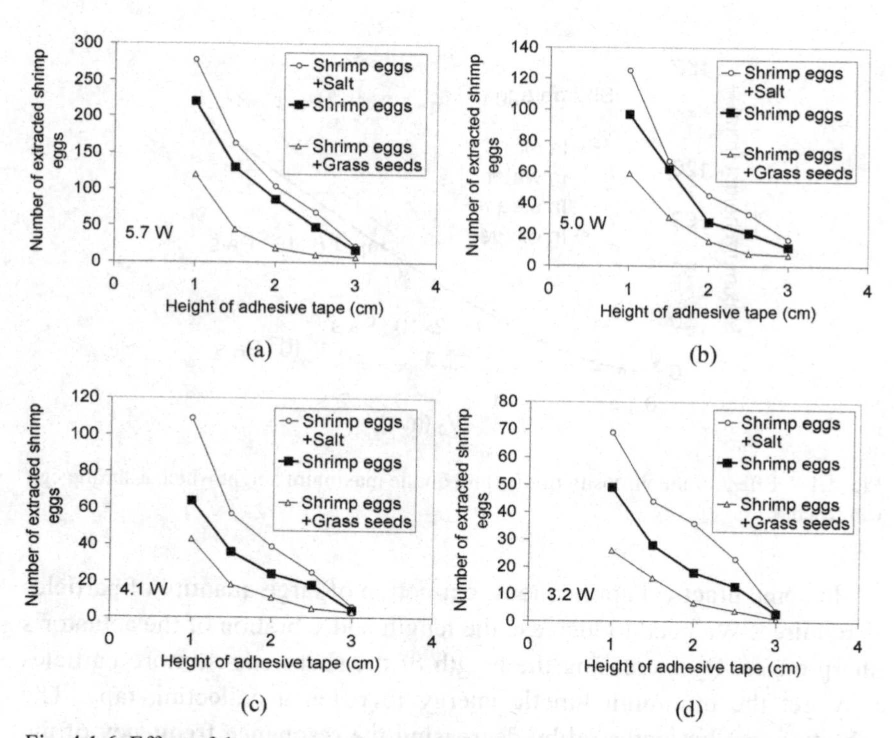

Fig. 4.1.6. Effect of the contents of a mixture on extraction capability at different input electric power (at resonance). (a) 5.7W; (b) 5.0W; (c) 4.1W; (d) 3.2W.

actuator. Other particles with larger mass, listed in Table I, were also tested in water. It was found that they could not be pumped up in water too. So the viscosity of the fluid cannot be ignored in analyzing the pumping effect.

To investigate the pumping effect under microgravity conditions, h_{max} versus g was investigated by using Eq. (4.1.22). Fig. 4.1.8(a) shows the relationship between h_{max} and gravitational acceleration g at different initial x-direction velocity V_0 of a shrimp egg in air. It is seen that the maximum height which particles can reach increases with decreasing the gravitational acceleration. Fig. 4.1.8(b) shows the relationship between h_{max} and g at $E_0 = 2\times10^{-9}$ J for shrimp eggs and grass seeds in air. It is seen that it is possible to extract lighter and smaller particles from a particle mixture under microgravity conditions.

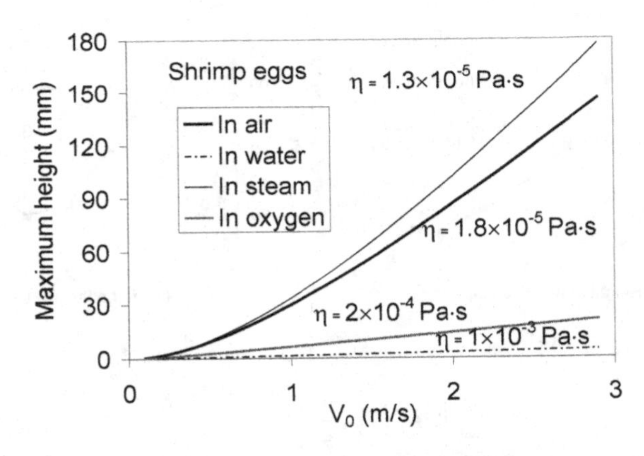

Fig. 4.1.7. Effect of the viscosity of the fluid on the maximum height which a shrimp egg can reach.

In some practical applications, extraction of large quantity of particles is required. We need to increase the length and vibration of the actuator's sharp edges. By increasing the length of the sharp edges, more particles may get the minimum kinetic energy to reach a collecting tape. The vibration can be increased by decreasing the resonance frequency of the actuator and optimizing the size and structure.

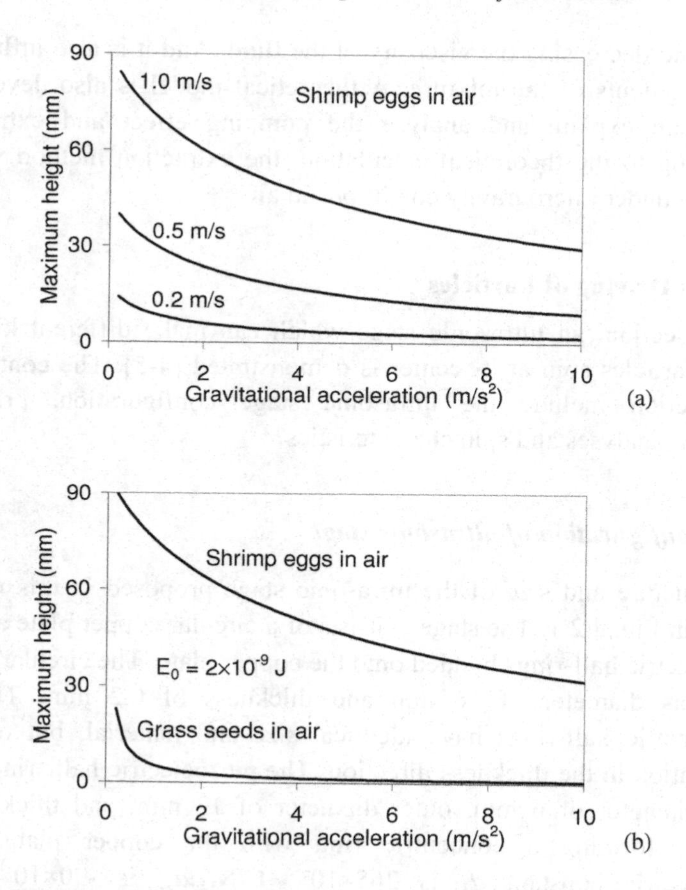

Fig. 4.1.8. Effect of the gravitation acceleration on the maximum height which a particle can reach in air. (a) The maximum height versus gravitation acceleration at different initial velocities of a shrimp egg; (b) the maximum height versus gravitation acceleration when of a shrimp egg and grass seed have constant initial kinetic energy .

4.1.5 *Summary*

A method of extracting particles from a particle mixture has been demonstrated. The particles with smaller mass and radius can be pumped up in the gap of a π-shaped actuator, and collected by an adhesive tape. Shrimp eggs can be effectively extracted from a mixture of shrimp eggs and fine salt crystals or grass seeds. The extraction capability increases with increasing the actuator's vibration, decreasing the adhesive tape height and decreasing the viscosity of the fluid. And it is also influenced by the contents of the mixture. A theoretical model is also developed, which can explain and analyze the pumping effect and extraction. According to the theoretical calculation, the extraction method will be effective under microgravity conditions in air.

4.2 Spin Driving of Particles

In this section, an ultrasonic stage which can make different kinds of micro particles spin at its center is demonstrated [4-5]. The contents of this section include the ultrasonic stage configuration, principle, vibration analyses and spin characteristics.

4.2.1 *Configuration of ultrasonic stage*

The structure and size of the ultrasonic stage proposed in this work is shown in Fig. 4.2.1. The stage consists of a circular copper plate and two piezoelectric half-rings bonded onto the copper plate. The circular copper plate has diameter of 16 mm and thickness of 0.2 mm. The two piezoelectric half-rings have identical size and material, but opposite polarization in the thickness direction. The piezoelectric half- rings have inner diameter of 6 mm, outer diameter of 12 mm, and thickness of 1.2 mm, forming a concentric ring with the copper plate. Their piezoelectric constant d_{33} is 265×10^{-12} C/N, d_{31} is -90×10^{-12} C/N, electromechanical coupling factor k_p is 0.5, mechanical quality factor Q_m is 800, dielectric dissipation factor $tan\delta$ is 0.5%, and density is 7700 kg/m^3. The experiments were conducted in air, and driving surface of the ultrasonic stage was perpendicular to the gravitation. During the experiments, the stage was clamped as shown in Fig. 4.2.1(b). From the bottom electrodes of piezoelectric half-rings to the copper plate, an AC voltage is applied to excite mechanical vibration in the copper plate. Measured resonance frequencies of the ultrasonic stage are 86.6 kHz and 93.76 kHz. The direction to observe the spin of micro particles is shown in Fig. 4.2.1(c).

Micro particles used in our experiment are shown in Fig. 4.2.2(a). The glass balls have two average diameters (0.51 mm and 0.75 mm); the shrimp eggs are spherical, and their average diameter is around 0.41 mm; the sweet alyssum seeds are ellipsoidal, and their average length and width are 1.66 mm and 1.17 mm, respectively; the grass seeds are approximately viewed as spherical, and their average diameter is 0.97 mm. The frictional coefficient between the particles and stage was measured as well as the mass of a single particle, and the results are shown in Table 4.2.1. The frictional coefficient was obtained by measuring the tilt angle of the stage when a measured particle starts to slip, and the mass per particle was obtained by measuring the total mass of several hundred particles and counting the exact number of measured particles.

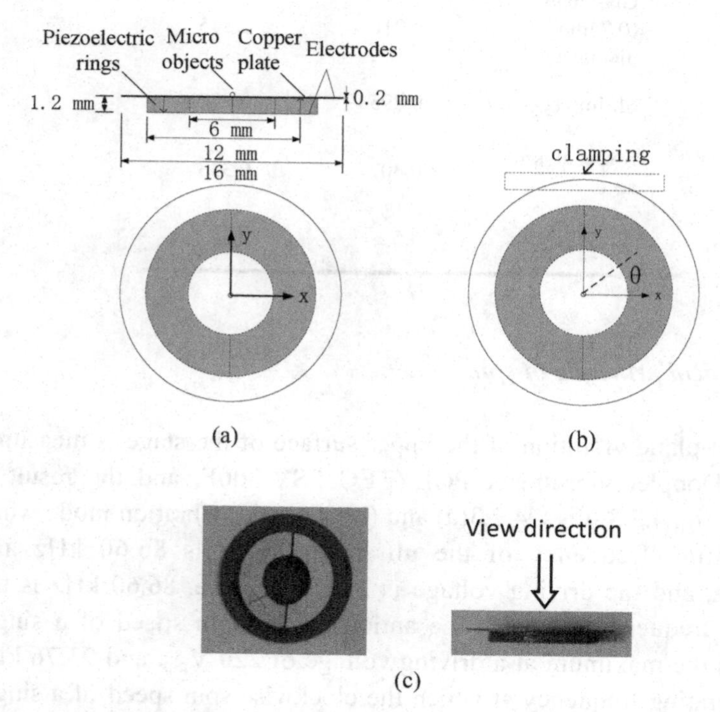

Fig. 4.2.1. The structure and size of the ultrasonic stage. (a) Schematic diagram and size of the ultrasonic stage. (b) Clamping of the ultrasonic stage. (c) Images of the ultrasonic stage.

It was observed that when the stage was in resonance, particles in the vicinity of the center of the stage ($x = 0$, $y = 0$) could move to the center due to the Chladni's effect, and spin at the center clockwise (around 93.76 kHz) or anticlockwise (around 86.6 kHz). In the Chladni's effect, particles in the vicinity of nodal points of a flexurally vibrating plate move to the nodal points. Fig. 4.2.2(b) shows the anticlockwise spin of a single glass seed at the center of the stage. It was taken by high speed camera Keyence VHX-1000E.

Table 4.2.1. Measured parameters of micro particles used in experiments.

Micro objects	friction factor	mass (mg)
Glass ball (0.75mm diameter)	0.216	0.545
Shrimp egg	0.295	0.024
Sweet alyssum seed	0.402	0.333
Grass seed	0.5	0.12

4.2.2 *Physical principle of spin*

The out-of-plane vibration of the upper surface of the stage is measured by laser Doppler vibrometer POLYTEC PSV-300F, and the result is shown in Fig. 4.2.3. Figs. 4.2.3(a) and (b) show the vibration mode when the operating frequency of the ultrasonic stage is 86.60 kHz and 93.76 kHz, and the driving voltage is 220 V_{p-p}. Here, 86.60 kHz is the operating frequency at which the anticlockwise spin speed of a single particle is the maximum at a driving voltage of 220 V_{p-p}, and 93.76 kHz is the operating frequency at which the clockwise spin speed of a single particle is the maximum at the same driving voltage. From this

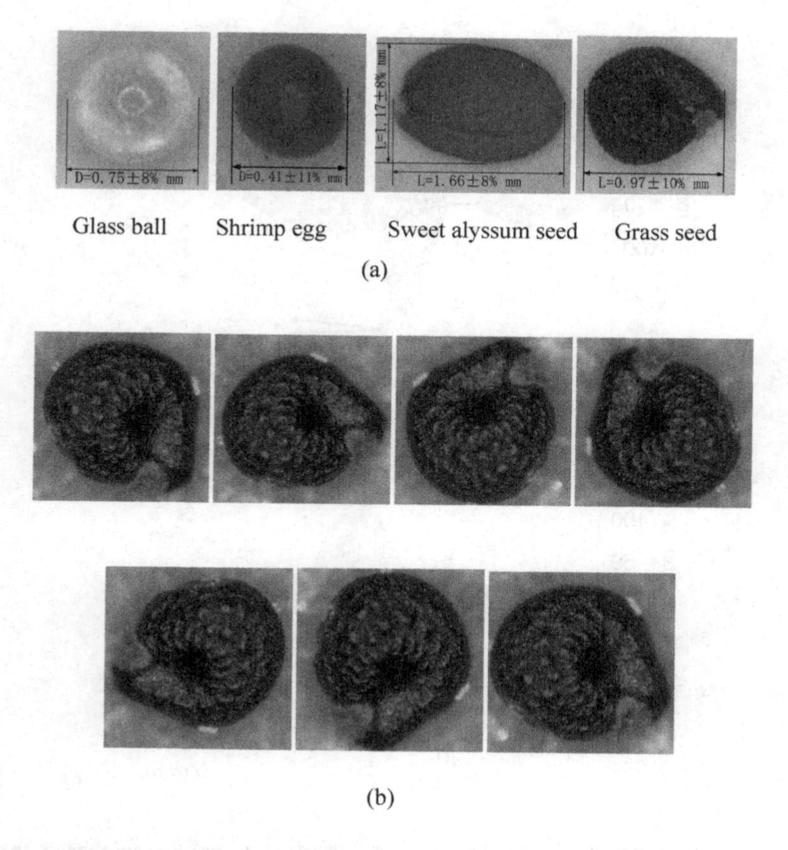

Fig. 4.2.2. Images of the micro particles used in the experiments. (a) Particles under microscope. (b) Grass seed in the anticlockwise spin.

measurement, it is known that there exists a nodal point at $x = 0$ and $y = 0$, which just locates at the center of the stage. The vibration mode of the stage is anti-symmetric about the center, and there are two out-of-phase peaks at the two sides of the nodal point. According to the measurement, the two peaks are at $x = \pm 0.9$ mm, $y = 0$ for 86.60 kHz, and $x = 0.636$ mm, $y = 0.636$ mm and $x = -0.636$ mm, $y = -0.636$ mm for 93.76 kHz.

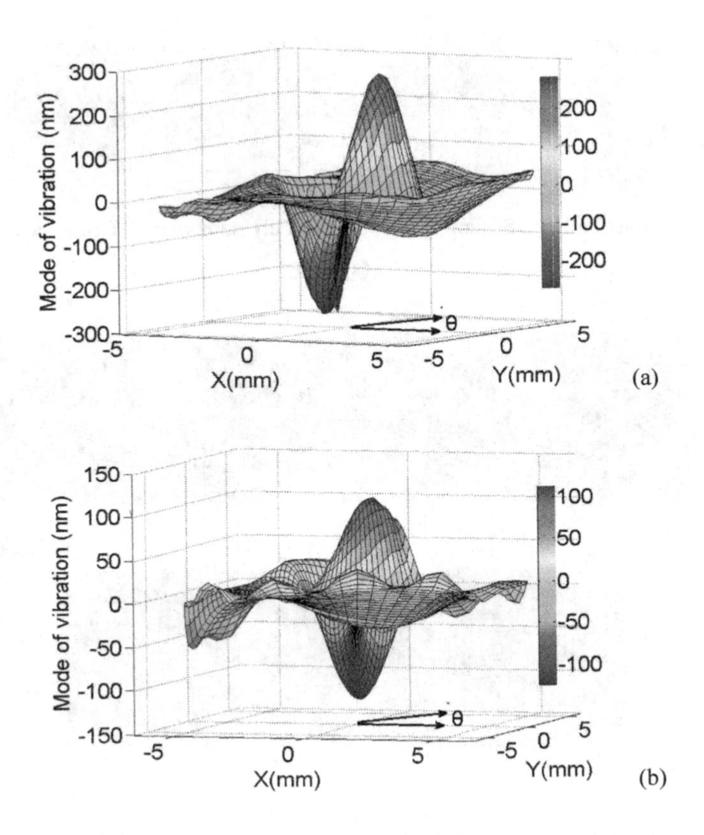

Fig. 4.2.3. Measured vibration displacement distribution of the ultrasonic stage. (a) At 86.6 kHz. (b) At 93.76 kHz.

Concentric circles with different radii R, centered at the nodal point, are drawn, and the amplitude and phase of the out-of-plane vibration displacement versus position angle θ on these circles is measured. The position angle θ is defined in Fig. 4.2.3. Figs. 4.2.4(a) and (b) show the results for R = 0.165 mm, 0.99 mm and 1.815 mm when the operating frequency is 86.6 kHz and the driving voltage is 220 $V_{p\text{-}p}$. It is seen that along the circles, the phase of vibration displacement increases as position angle θ increases; in regions A, B, C and D, the phase versus position angle θ is approximately linear although it has different slopes in these regions. Figs. 4.2.5(a) and (b) show the results when the

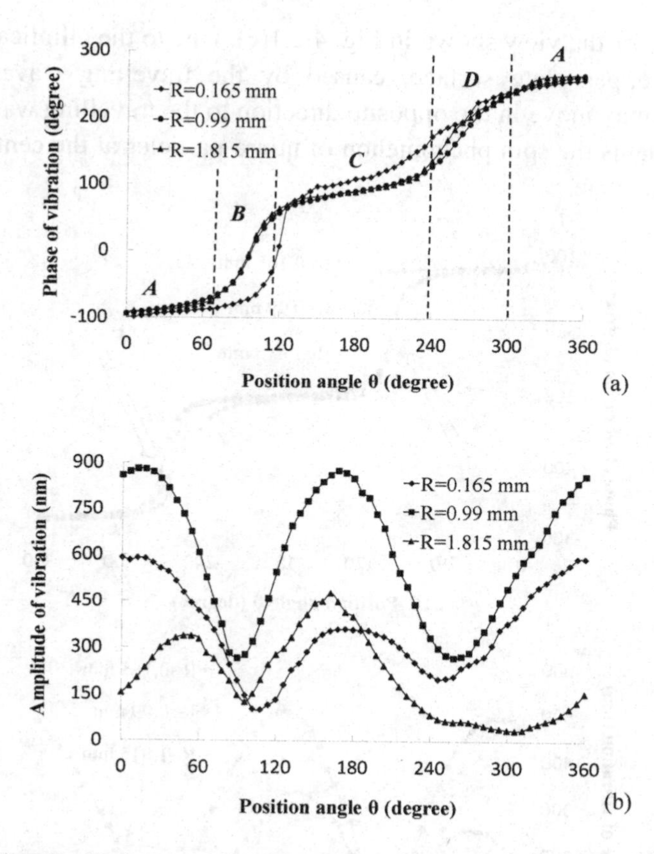

Fig. 4.2.4. Vibration displacement distribution around the central point of the ultrasonic stage at operating frequency of 86.6 kHz. (a) Phase. (b) Amplitude.

operating frequency and voltage are 93.76 kHz and 220 V_{p-p}, respectively. It is seen that in regions *E* and *F*, the phase decreases as position angle θ increases, and the phase versus position angle is approximately linear. Therefore, at 86.6 kHz and 93.76 kHz, there are travelling waves around the nodal point (the center of the stage). At 86.6 kHz, there are travelling waves in regions *A*, *B*, *C*, and *D*, and they are in the $-\theta$ direction in Fig. 4.2.3 or clockwise direction for the view shown in Fig. 4.2.1(c); at 93.67 kHz, there are travelling waves in regions *E* and *F*, and they are in the $+\theta$ direction or anticlockwise direction for the view shown in Fig. 4.2.1(c). Due to the elliptical motion of the copper plate surface, caused by the travelling waves, micro particles may move in the opposite direction to the travelling waves. This well explains the spin phenomenon of micro particles at the center of the ultrasonic stage.

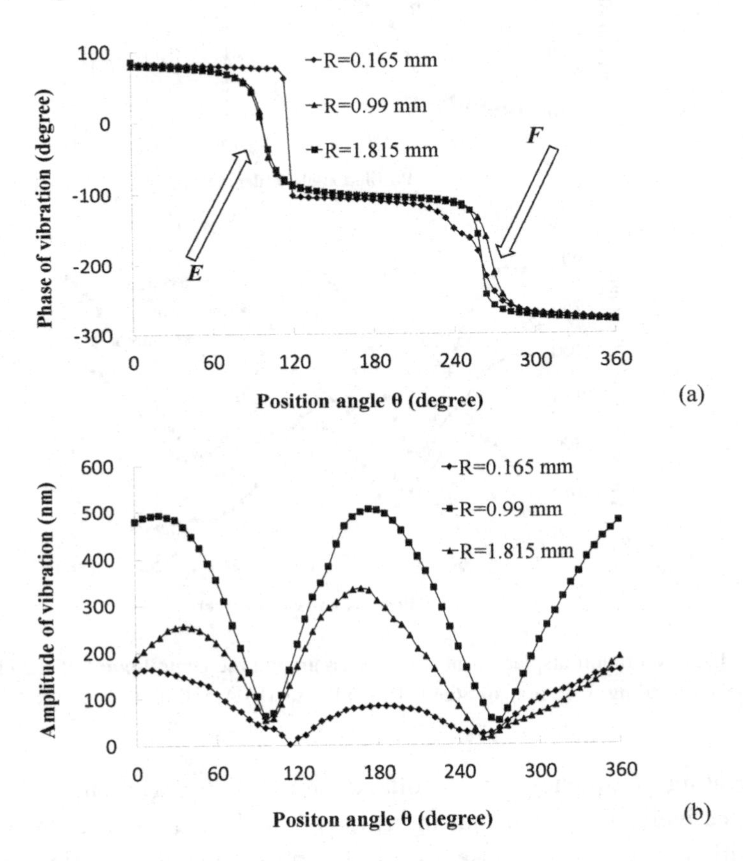

Fig. 4.2.5. Vibration displacement distribution around the central point of the ultrasonic stage at operating frequency of 93.76 kHz. (a) Phase. (b) Amplitude.

4.2.3 *Vibration analyses and spin characteristics*

Vibration of the stage was calculated by the FEM software ANSYS to investigate the vibration pattern of the piezoelectric half-rings. Fig. 4.2.6

shows a 3D mesh model used in the calculation. The size and material of the model are the same as those of the ultrasonic stage described in section 4.2.1. The solid5 elements were used for the piezoelectric half-rings and solid45 elements were used for the circular copper plate; the β damping ratio ($=7.7\times10^{-8}$ 1/Hz) and sparse solver were used for the harmonic response calculation; the operating frequency was 86.6 kHz and 93.76 kHz. Fig. 4.2.7 shows the calculated vibration displacement of the ultrasonic stage at 86.6 kHz. It is seen that the two piezoelectric half-rings vibrate extensionally in the circumferential direction; when one piezoelectric half-ring extends, another one contracts. According to the calculation, the two out-of-phase vibration peaks are at $x = -0.95$ mm, $y = 0$ and $x = 0.90$ mm, $y = 0$, respectively, which agrees with the measured result shown in Fig. 4.2.3(a) quite well.

Fig. 4.2.8 shows the measured anticlockwise spin speed vs. operating frequency for a single glass ball (the average diameter = 0.747 mm), shrimp egg, sweet alyssum seed and grass seed at 140 $V_{p\text{-}p}$ driving voltage. The spin speed was measured by slowing down the video clips of the particle spin, and counting the revolution number of the spin within a given time. Video clips of the particle spin were taken by Cannon EOS 550D camera. It is seen that different particles have different maximum spin speed at resonance, and the shrimp egg has the maximum spin speed at resonance among the four particles.

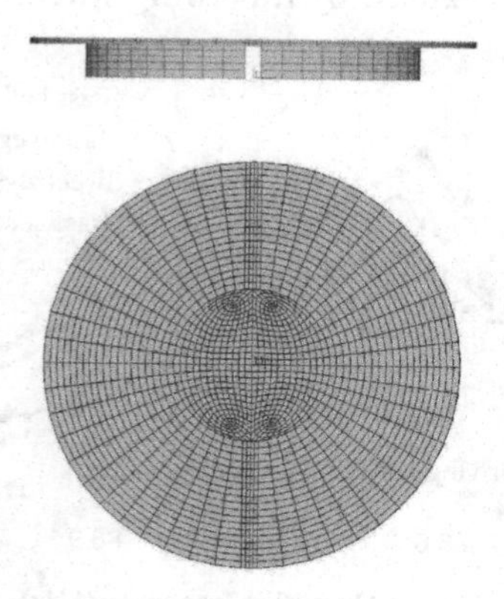

Fig. 4.2.6. Meshed FEM model of the ultrasonic stage.

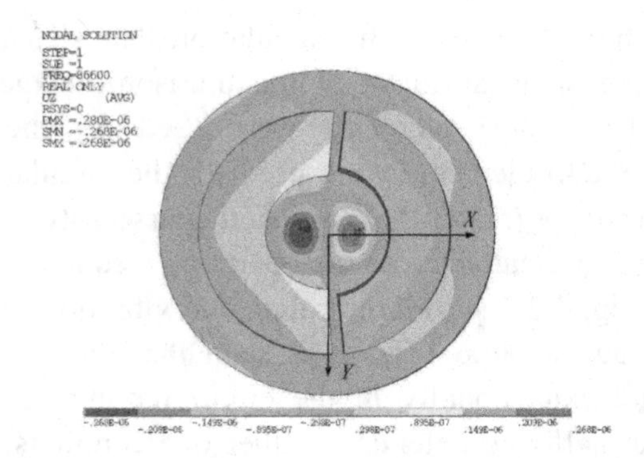

Fig. 4.2.7. FEM results of the vibration displacement distribution of the ultrasonic stage.

The anticlockwise spin speed vs. vibration displacement amplitude for a single glass ball with a diameter of 0.51 mm and 0.75 mm was measured around 86.6 kHz, and the result is shown in Fig. 4.2.9. The vibration measurement is at point x = 0.9 mm, y = 0 mm on the upper surface of the stage, which is also the peak vibration point of the stage. It is seen that the glass ball with 0.51 mm diameter has higher spin speed than that with 0.75 mm diameter for the whole vibration displacement amplitude range, and the larger the amplitude of vibration displacement is, the higher the spin speed is. The relationship between the spin speed and vibration displacement amplitude is quite linear, which supports the principle that the spin is caused by travelling wave.

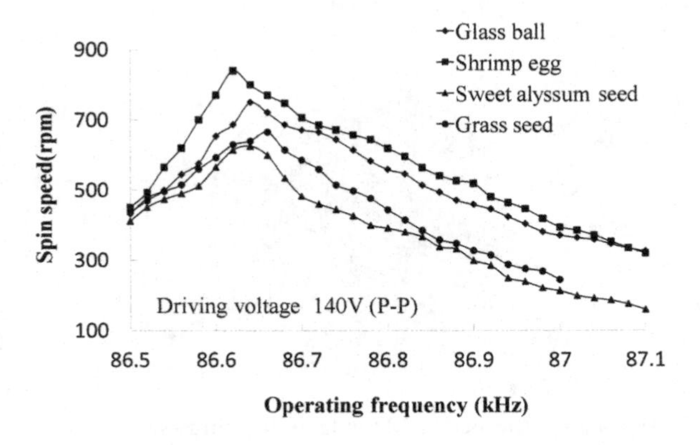

Fig. 4.2.8. Spin speed vs. operating frequency for different kinds of micro particles.

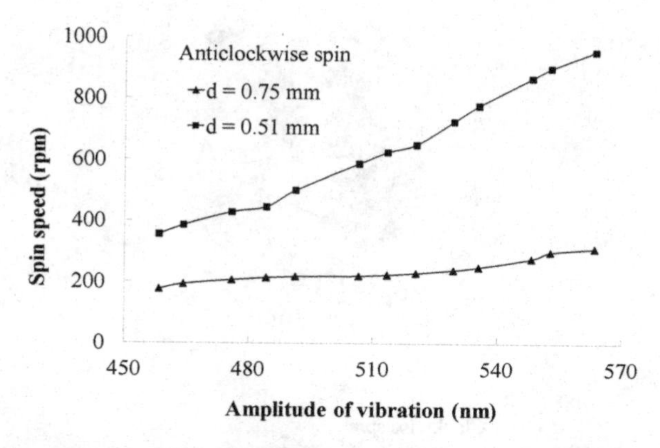

Fig. 4.2.9. Spin speed vs. vibration displacement amplitude (0-peak) for the glass balls with diameter of 0.51 mm and 0.75 mm.

It was observed that a combination of multiple particles could also spin around the center of the stage. Fig. 4.2.10(a) shows the spin of a two particle system. Shrimp egg *A* is at the center of the stage, and shrimp egg *B* sticks on the surface of shrimp egg *A*. The spin center is in shrimp egg *A*. The spin speed vs. operating frequency around 93.7 kHz and 86.6 kHz was measured and the result is shown in Figs. 4.2.10(b) and (c). For the same operating voltage, the maximum speed of the anticlockwise spin is higher than that of the clockwise spin. This is because the vibration displacement of the stage in resonance decreases as the resonance frequency increases for the same driving voltage.

4.2.4 *Summary*

An ultrasonic stage which can make different kinds of micro particles spin at its center is demonstrated. In the stage, a circular copper plate is mechanically excited by two piezoelectric half-rings bonded onto the bottom surface of the copper plate, and the circumferential extensional vibration of the piezoelectric half-rings are employed. The spin of micro particles is caused by travelling waves around the center of the stage. A single particle or combination of several particles can spin around the

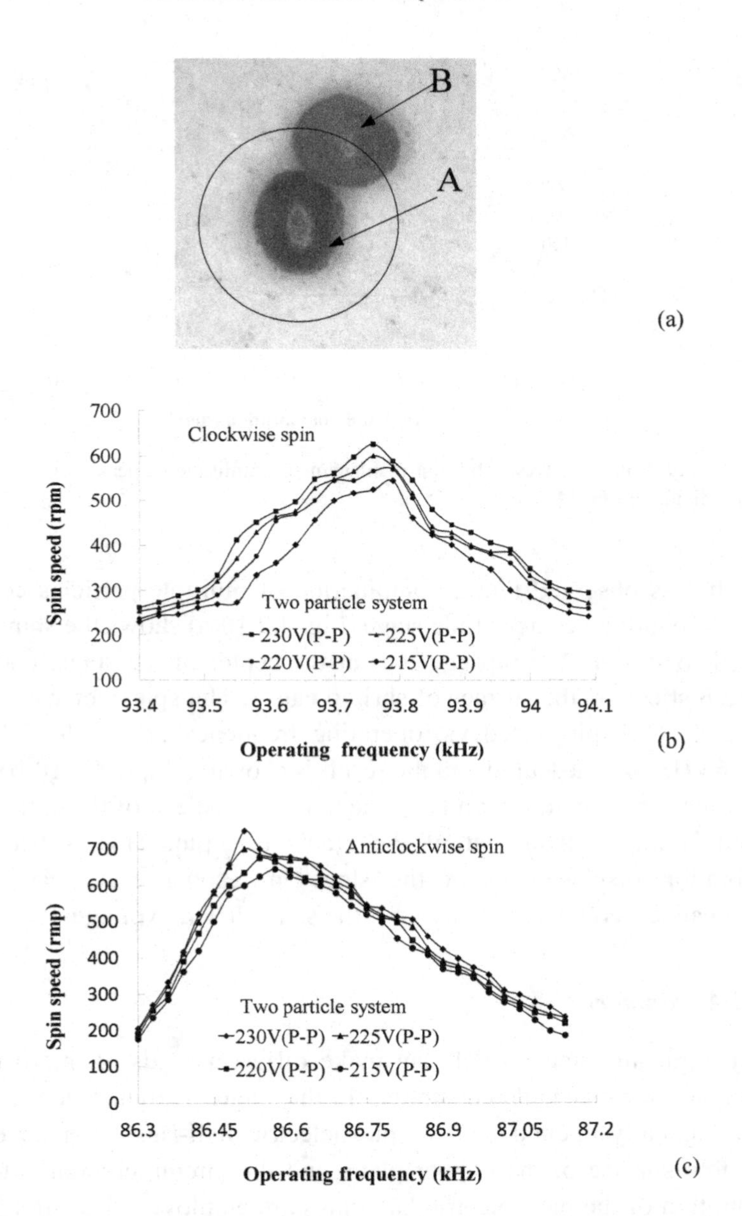

Fig. 4.2.10. Spin speed vs. operating frequency for the two particle system of shrimp eggs. (a) Configuration of the two particle system. (b) Clockwise spin. (c) Anticlockwise spin.

center of the stage. The particles used in the experiments include glass balls with diameter of 0.75 mm and 0.51 mm, spherical shrimp eggs with diameter of 0.41 mm, ellipsoidal sweet alyssum seeds 1.66 mm long and 1.17 mm wide, and grass seeds with diameter of 0.97 mm. The spin direction can be reversed by using different operating frequency. The spin speed can be controlled by the operating frequency and driving voltage of the stage, and it reaches 955 rpm for a single glass ball with 0.51 mm diameter. The rotary drive of micro particles on an ultrasonic substrate has potential applications in the measurement of cohesive force between particles, observation and measurement of the flow caused by a spinning particle with complex shape, observation and measurement of sound induced eddy flow in a droplet, separation of different micro particles, etc.

4.3 Rotary Driving of Small Mechanical Components

Rotary driving of small mechanical components by the travelling wave around a nodal point of vibration is demonstrated in this section [6]. The

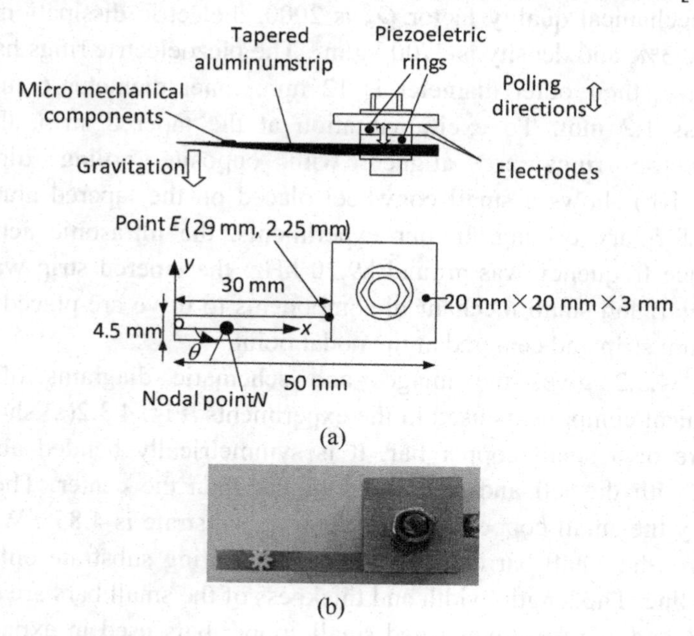

Fig. 4.3.1. The structure and size of the ultrasonic actuator. (a) Schematic diagram and size of the actuator. (b) A small cogwheel for rotary drive.

contents of this section include an experimental setup for showing the rotary driving phenomenon, principle analyses, characteristics and discussions.

4.3.1 *Experimental setup and phenomena*

Fig. 4.3.1(a) shows the structure and size of the ultrasonic actuator used in our experiments. The tapered aluminum strip is 4.5 mm wide and 30 mm long, which is mechanically driven by a sandwich type ultrasonic actuator. The thickness of the tapered strip is 3 mm at its root and decreases to 0.3 mm linearly as approaching the sharp end of the tapered strip. By using this tapered structure, large vibration up to several microns (zero-peak) can be obtained in the sharp end of the metal strip. In the ultrasonic actuator, two piezoelectric rings (Fuji C203) are pressed against each other between two square aluminum plates by a bolt structure. The aluminum plates are 3 mm thick each. The piezoelectric constant d_{33} is 325×10^{-12} C/N, electromechanical coupling factor k_{33} is 0.71, mechanical quality factor Q_m is 2000, dielectric dissipation factor $\tan\delta$ is 0.3%, and density is 7700 kg/m^3. The piezoelectric rings have the same size; their outer diameter is 12 mm, inner diameter 6 mm and thickness 1.2 mm. To excite vibration at the tapered strip, the two piezoelectric rings are aligned with opposite poling direction. Fig. 4.3.1(b) shows a small cogwheel placed on the tapered aluminum strip for rotary driving. In our experiments, the ultrasonic actuator's resonance frequency was around 89.20 kHz; the tapered strip was kept horizontal; and small mechanical components to drive are placed on the aluminum strip and centered at the nodal point.

Fig. 4.3.2 gives the images and schematic diagrams of small mechanical components used in the experiments. Fig. 4.3.2(a) shows the structure of a small copper bar. It is symmetrically bended about its center, with the left and right sides higher than the center. The angle made by the small copper bar and vibrating substrate is 4.85°. With this structure, the small bar contacts with the vibrating substrate only at its central line. The length, width and thickness of the small bars are defined as *L*, *W* and *T*, respectively, and small copper bars used in experiments have the same thickness and width but different length. They are 0.1 mm thick, 1 mm wide, and 3 mm, 4 mm, 5 mm, 6mm and 7 mm long, respectively. Fig. 4.3.2(b) shows the images and schematic diagrams of composite structures of the small copper bars. Each copper bar is 7 mm long, 1 mm wide and 0.1 mm thick. The small copper bar in contact with the tapered aluminum strip is also bended in the same way as shown in Fig. 4.3.2(a). Fig.4.3.2(c) shows the image and schematic diagram of a small copper disk, which are also bended symmetrically about the center. Its diameter and thickness are denoted by d and T, respectively. The angle made by the disk surface and vibrating strip is 8.9°. Fig. 4.3.2(d) shows the image and schematic diagram of a small plastic cogwheel used in our experiments. The inner and tip diameter of the cogwheel is 1 mm and 5 mm, respectively, the teeth number is 8, and the cogwheel height is 4 mm. On its side surface contacting with the vibrating aluminum strip, there is a concentric plastic projection ring with a central diameter of 1.65 mm, height of 0.2 mm, and width of 0.65 mm. Table 4.3.1 lists the frictional coefficient, density and mass of the small mechanical components used in our experiments. The mass of the small bar composites can be calculated from the data in the table.

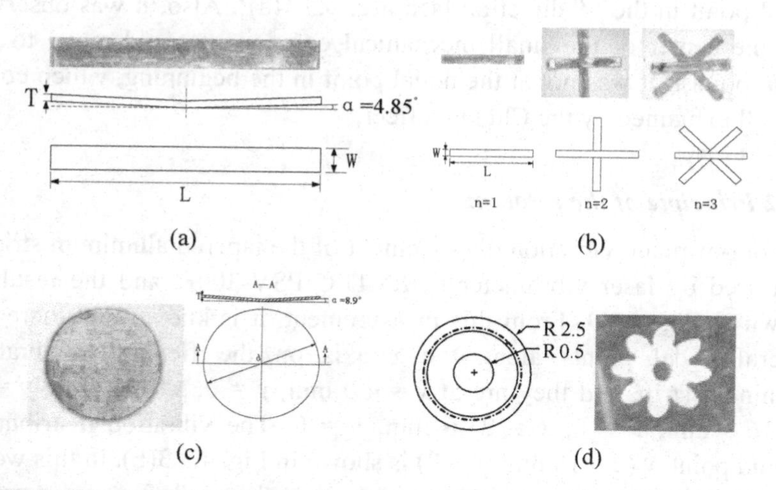

Fig. 4.3.2. Images and schematic diagrams of the small mechanical components. (a) Small copper bar. (b) Composites of small copper bars. (c) Small copper disk. (d) Small cogwheel.

Table 4.3.1. Frictional coefficient, density and mass of the mechanical components.

Mechanical components	Coefficient of friction	Density (g/cm^3)	Mass (g)
Small copper bar	0.01	8.5	0.0025 (L=3 mm); 0.003 (L=4 mm); 0.004 (L=5 mm); 0.005 (L=6 mm); 0.006 (L=7 mm)
Small copper disk	0.01	8.5	0.014 (d=4.7 mm T=0.1 mm); 0.022 (d=5.7 mm T=0.1 mm); 0.017 (d=3.6 mm T=0.2 mm)
Small cogwheel	0.014	0.94	0.053

In the experiments, it was observed that when the ultrasonic actuator was in resonance (~89.20kHz), there was a vibration nodal point at x = 10 mm, y = 0 on the surface of the tapered aluminum strip and the small mechanical components centered at this point could rotate around this nodal point in the $-\theta$ direction [see Fig. 4.3.1(a)]. Also, it was observed that the center of the small mechanical component could move to the nodal point if it was not at the nodal point in the beginning, which could be well explained by the Chladni effect.

4.3.2 *Principle of the rotation*

The out-of-plane vibration displacement of the tapered aluminum strip is measured by laser vibrometer POLYTEC PSV-300F, and the result is shown in Fig. 4.3.3. From this measurement, it is known that there are several nodal points along the x axis on the flexurally vibrating aluminum strip, and they are at x = 4.0 mm, y = 0; x = 10 mm, y = 0; x = 16.8 mm, y = 0; x = 23.6 mm, y = 0. The vibration distribution around point N (x = 10 mm, y = 0) is shown in Fig. 4.3.3(b). In this work, only nodal point N (x = 10 mm, y = 0) is used for our detailed investigation.

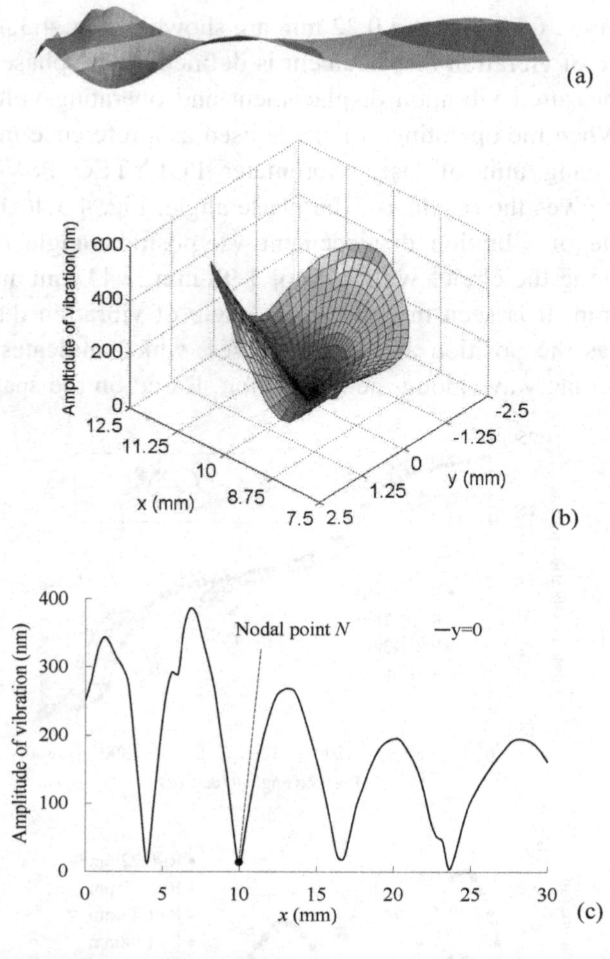

Fig. 4.3.3 Measured vibration displacement distribution of the aluminum strip. (a) Mode shape. (b) Vibration distribution around a nodal point N at x = 10 mm, y = 0. (c) The out-of-plane vibration displacement distribution along the x-axis.

It is well known that one of the operation principles of ultrasonic motors is to use a travelling wave to drive a rotor. To understand whether the actuator in our experiment has the similar driving principle, concentric circles with different radii R and centered at nodal point N were drawn, and the phase angle and amplitude of vibration displacement on these circles were measured. The results for R = 1.98 mm, 1.43 mm , 0.77 mm and 0.22 mm are shown in Fig. 4.3.4. Here, the phase angle of vibration displacement is defined by the phase difference between measured vibration displacement and operating voltage of the actuator. When the operating voltage is used as a reference input for the data processing unit of laser vibrometer POLYTEC PSV-300F, the vibrometer gives the reading of the phase angle. Fig. 4.3.4(a) shows the phase angle of vibration displacement vs. position angle θ [see Fig. 4.3.1(a)] along the circles with radii of 1.98 mm, 1.43 mm and 0.77mm and 0.22 mm. It is seen that the phase angle of vibration displacement decreases as the position angle θ increases, which indicates that there exists travelling wave along the θ direction. Based on the spatial change

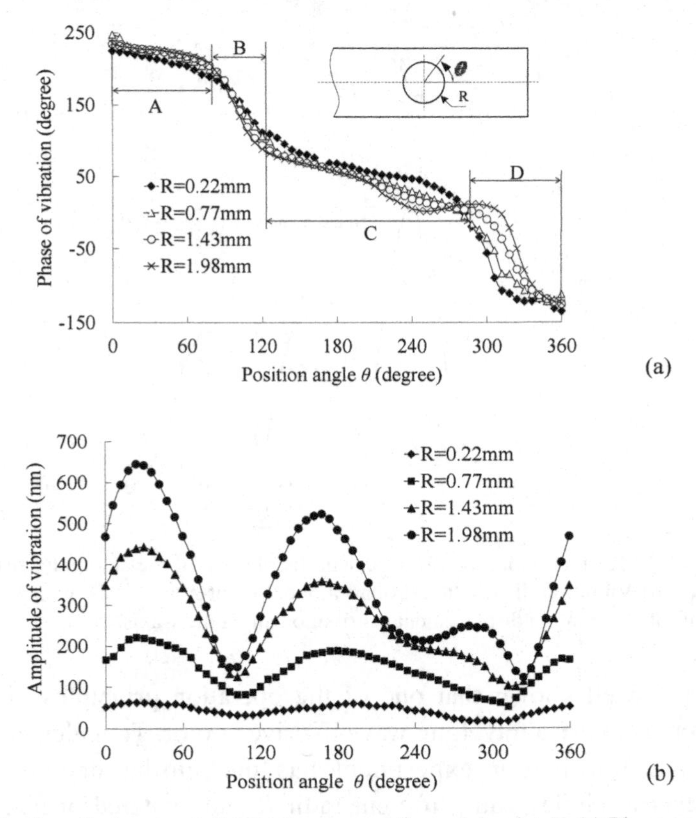

Fig. 4.3.4. Vibration displacement distribution around the nodal point N. (a) Phase angle. (b) Amplitude.

rate of the phase angle, the concentric circles around nodal point *N* can be divided into four sections *A, B, C* and *D*, in which the travelling waves have different speed but identical direction (the $+\theta$ direction). Therefore, the rotation of small mechanical components around nodal point *N* may be caused by elliptical motion of the surface of the tapered aluminum strip, resulting from the travelling waves. The observed rotation direction (the $-\theta$ direction), which is opposite to the traveling wave direction, supports this conclusion.

4.3.3 *Experimental results and discussion*

In the following experiments, unless otherwise specified, the revolution speed of small mechanical components was measured in air; the revolution speed was obtained by slowing video clips of the rotation and counting the revolution number in one second; the vibration displacement was measured by laser vibrometer POLYTEC PSV-300F; the out-of-plane vibration at point *E* (x = 29 mm, y = 2.25 mm) (see Fig. 4.3.1(a)) was used to represent vibration strength of the tapered metal strip.

4.3.3.1 *Characteristics of small copper bars and their composites*

Fig. 4.3.5(a) shows the measured revolution speed vs. operating frequency for the single small copper bars with different lengths but identical width and thickness at a driving voltage of 120 Vp-p. It is seen that the revolution speed reaches maximum near the resonance point. Fig. 4.3.5(b) shows the measured maximum revolution speed vs. the length of the small copper bar at driving voltage of 120 Vp-p and operating frequency of 89.2 kHz. It is seen that the maximum revolution speed decreases as the length of the small copper bar increases. The possible reason for this phenomenon is that as the weight of the small copper bar increases, the frictional force on the rotating bar increases. In Fig. 4.3.5(b), the maximum revolution speed reaches 630 rpm. Fig.4.3.6 shows the measured revolution speed vs. vibration displacement amplitude (0-peak) for the single small copper bars. The revolution speed increases as the vibration displacement amplitude (0-peak) increases,

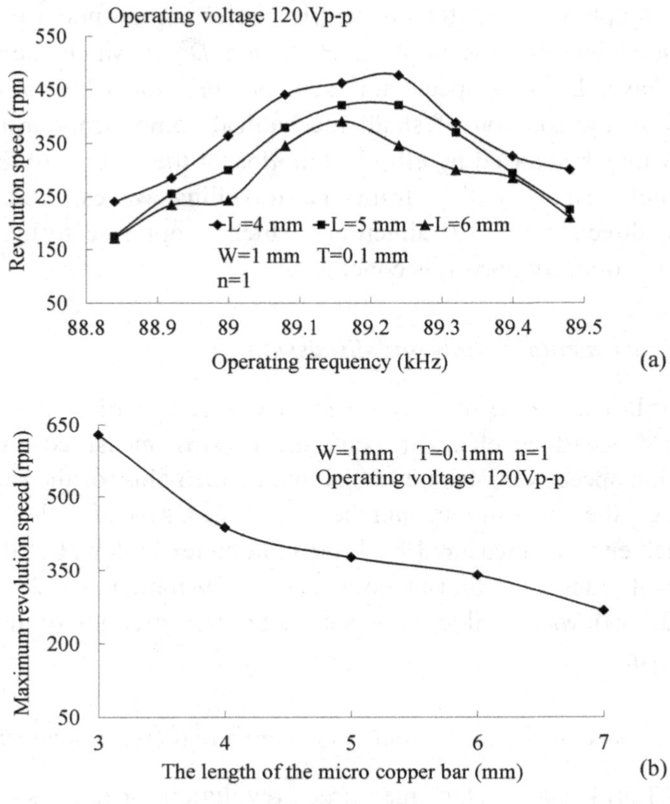

Fig. 4.3.5. Revolution speed characteristics of small single copper bars. (a) Revolution speed vs. operating frequency at driving voltage of 120Vp-p. (b) Revolution speed vs. the length of small single copper bars.

which is caused by the increase of amplitude of the travelling waves around nodal point *N*.

Fig. 4.3.7 shows the measured revolution speed vs. operating frequency for the small copper bar composites shown in Fig. 4.3.2(b), at driving voltage of 120 $V_{p\text{-}p}$. From this figure, it is found that the single small copper bar has the maximum revolution speed much higher than either of the composite structures, and the maximum revolution speed of the composite formed by three small copper bars is slightly lower than that formed by two small copper bars. This is because the frictional force on a composite structure increases as the number of small copper bars in the composite structure increases. The revolution speed vs. operating frequency was also measured for the single copper bar in vacuum (0.15 atm), and the result is also shown in Fig. 4.3.7. It is seen that the difference of revolution speed in air and vacuum is very little, which indicates that the effect of air resistance on the revolution is very small.

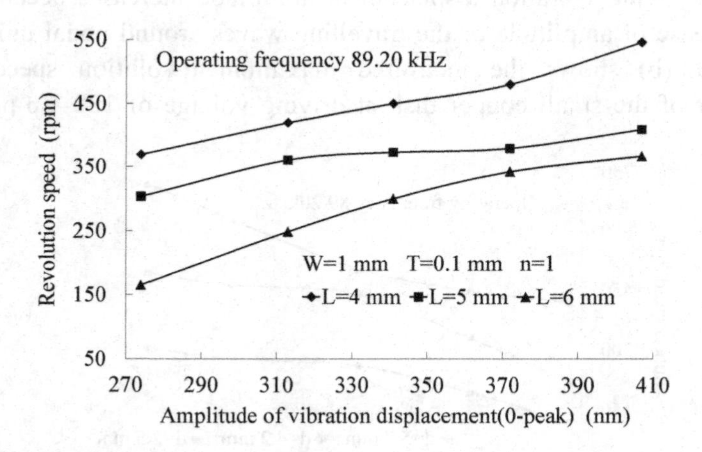

Fig. 4.3.6. Revolution speed vs. the vibration displacement amplitude (0-peak) for the small single copper bars.

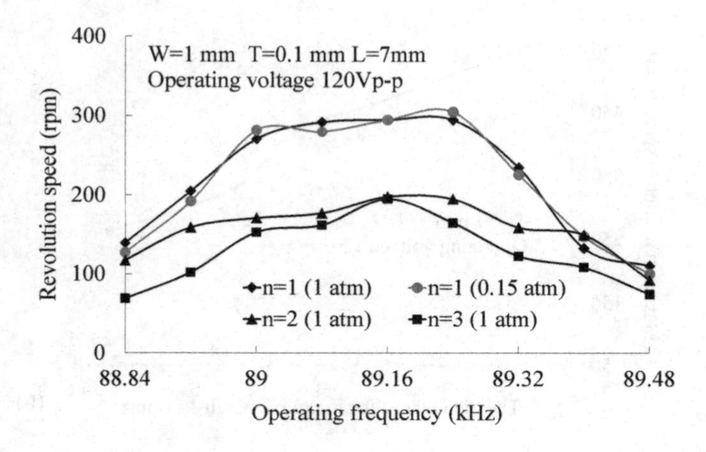

Fig. 4.3.7. Revolution speed vs. operating frequency for the small copper bar composites at driving voltage of 120 $V_{p\text{-}p}$.

4.3.3.2 *Characteristics of small copper disks and cogwheel*

Fig. 4.3.8(a) shows the measured revolution speed vs. vibration displacement amplitude (0-peak) for the small copper disks with different diameters but identical thickness. The revolution speed increases as the vibration displacement amplitude increases because of the increase of amplitude of the travelling waves around nodal point *N*. Fig. 4.3.8(b) shows the measured maximum revolution speed vs. diameter of the small copper disk at driving voltage of 120 Vp-p. The

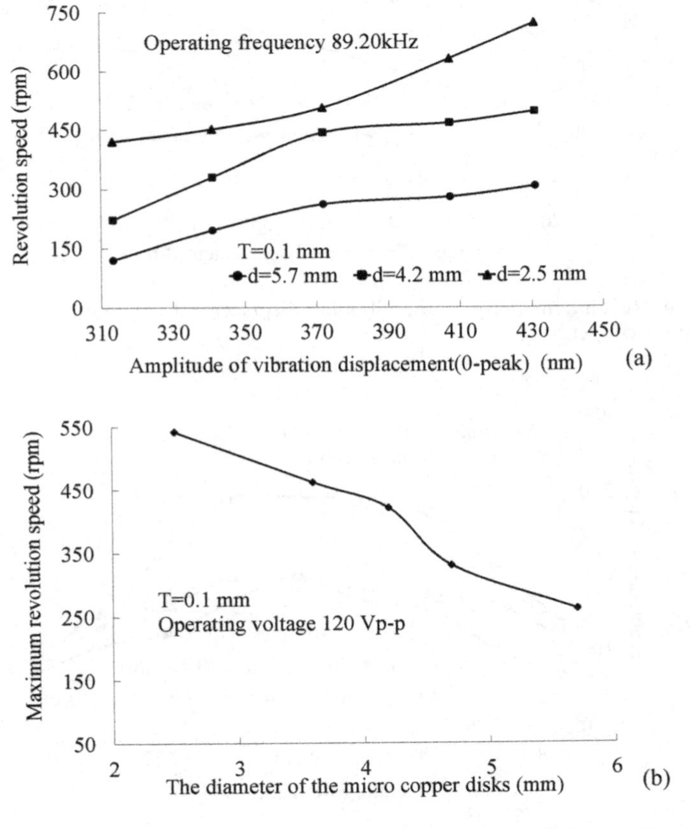

Fig. 4.3.8. Revolution speed characteristics of the small copper disks. (a) Revolution speed vs. the vibration displacement amplitude (0-peak) at point *E*. (b) Revolution speed vs. the copper disk diameter.

smaller the disk diameter, the larger the revolution speed is. This is because as the diameter of the disk becomes small, the frictional force on the disk also becomes small. Fig. 4.3.9 shows the measured maximum revolution speed vs. thickness of the small copper disk with a diameter of 3.6 mm at driving voltage of 120 Vp-p. It is seen that the revolution speed decreases as the thickness of the small copper disk increases. This phenomenon should be caused by the frictional force increase with thickness increase of the micro copper disk.

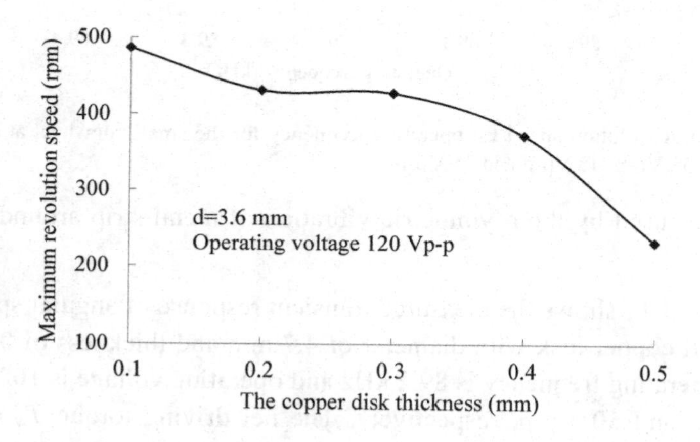

Fig. 4.3.9. Revolution speed vs. the copper disk thickness.

In the above described experiments, the small mechanical components were driven by the surface of the metal strip adjacent to nodal point *N*. Fig. 4.3.10 shows the measured revolution speed vs. operating frequency for the small cogwheel shown in Fig. 4.3.2(d). In the experiment, the cogwheel axis was perpendicular to the vibrating metal strip, and a projection ring on the side of the cogwheel was in contact with the metal strip in vibration; thus the cogwheel was driven by a circular strip surface centered at nodal point *N* on the metal strip. It was observed that the center of the cogwheel moved away from nodal point *N* when operating voltage was larger than 55 Vp-p approximately with operating frequency close to the resonance frequency. This phenomenon

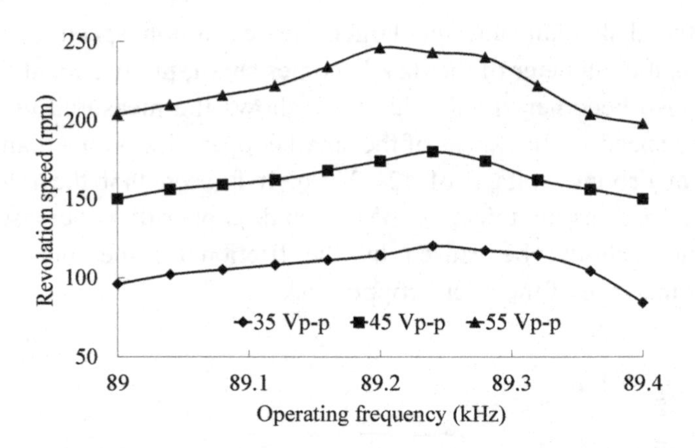

Fig. 4.3.10. Revolution speed vs. operating frequency for the small cogwheel at driving voltage of 55 Vp-p, 45 Vp-p and 35 Vp-p.

may be caused by the asymmetric vibration of metal strip around nodal point *N*.

Fig. 4.3.11 shows the measured transient response of angular speed of the small copper disk with diameter of 4.7 mm and thickness of 0.1 mm when operating frequency is 89.2 kHz and operating voltage is 100 Vp-p, 80 Vp-p and 50 Vp-p, respectively. The net driving torque T_N can be calculated by

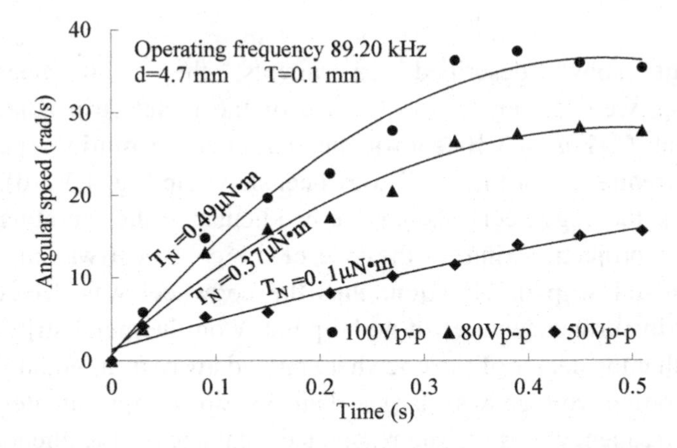

Fig. 4.3.11. Transient response of angular speed of the small copper disk with 4.7 mm diameter and 0.1 mm thickness at the starting.

$$T_N = J\frac{d\omega}{dt} \tag{4.3.1}$$

where J is the inertia moment of the small copper disk and ω the angular speed. The inertia moment can be calculated by the following equation:

$$J = \frac{1}{2}mr^2 \tag{4.3.2}$$

where m (= 0.014g) and r are the mass and radius of the small copper disk, respectively. By equations (4.3.1) and (4.3.2), we calculated the net driving torque on the small copper disk. For the curves in Fig. 4.3.11, the net driving torque is 0.49 μNm, 0.37 μNm and 0.1 μNm, respectively.

Angular speed versus time was measured after switching off the driving voltage, and the result is shown in Fig. 4.3.12. The frictional torque is estimated from the temporal change rate of angular speed at low angular speed range (just before stop), and it is 0.42 μNm approximately. Therefore the driving torque applied by the vibrating metal strip on the disks for the three curves in Fig. 4.3.11 is 0.91 μNm, 0.79 μNm and 0.52 μNm, respectively. Near resonance, the driving torque increases as operating voltage increases.

4.3.4 *Summary*

It is demonstrated that that small mechanical components such as bars, bar composites, disks and cogwheels, centered at a nodal point of a tapered metal strip in flexural vibration, can be driven to rotate. The revolution speed can be increased by increasing the tapered metal strip's vibration and decreasing the mechanical components' size. Driving torque on a disk with 4.7 mm diameter and 0.1 mm thickness is estimated to be 0.91 μNm when the operating voltage is 100 Vp-p and the device is in resonance. Travelling waves around the nodal point cause the rotary drive. The method presented here provides a new mechanism for rotary drive of small mechanical components.

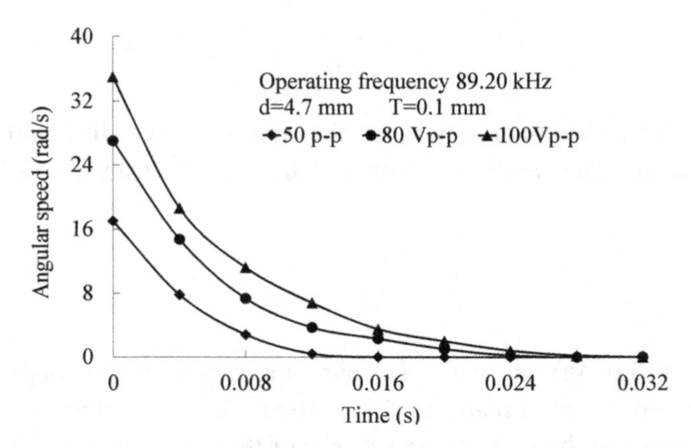

Fig. 4.3.12. Transient response of angular speed of the small copper disk with 4.7 mm diameter and 0.1 mm thickness after switching off the driving voltage.

4.4 Dust Removal for Photovoltaic Panels

Removing dust from the surface of a solar panel is a good example of large-scale ultrasonic manipulations. This section demonstrates a linear piezoelectric actuator based solar panel cleaning system [7], which can effectively and energy efficiently clean the active surface of a solar panel. The contents of this section include the structure and principle of the linear piezoelectric actuator, characteristics of the actuator, construction and energy analyses of the cleaning system, characteristics and discussion.

4.4.1 *Linear piezoelectric actuator and assembling*

4.4.1.1 *Structural design*

The linear piezoelectric actuator consists of two vibrators with identical structure and materials and a support. As shown in Fig. 4.4.1, each vibrator has two driving feet (*A* & *B*) with the same structure and materials. The two driving feet are assembled face to face to a linear guide. Each vibrator consists of driving bar *1*, electrodes *2*, piezoelectric plates *3* and *4*, and fastening screw *5*. Each piezoelectric plate (10 mm × 10 mm × 1mm) has two conversely polarized areas, and piezoelectric plates *4* rotate 90°around the central axial relatively to the position of piezoelectric plates *3*. Electrodes and piezoelectric plates are stacked between the driving bar and support, fastened by a screw. Piezoelectric plates *3* are used to excite the bending vibration mode in one of the driving feet in the *z* direction, and piezoelectric plates *4* to excite another bending vibration mode in the *y* direction in the same driving foot. Thus two motions of the tip of the driving feet are orthogonal to each other. In order to adjust frequencies of the two modes to be the same, slots are cut out on the driving bars. Driving feet *A* and *B* are used to form two symmetric elliptical motions for the linear driving. The reason for using the actuator topology shown in Fig. 4.4.1 is to obtain sufficient driving force with a compact structure.

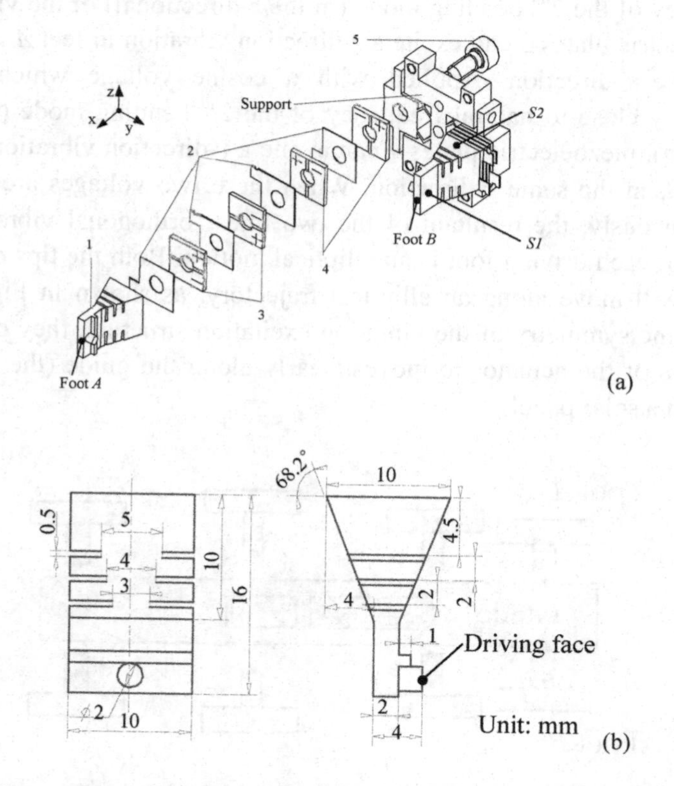

Fig. 4.4.1. Construction of the vibrator. (a) Components and assembling. (b) Size of the driving bar.

The apparent size of the two vibrators of the actuator is 32.6mm× 30mm×26.8mm, and more details about the driving bars' size parameters are shown in Fig. 4.4.1(b). Piezoelectric material used in the vibrators is PZT-8H. It has the density of 7450 kg/m^3, electromechanical coupling factor k_{33} of 0.60, piezoelectric constant d_{33} of 200×10^{-12} C/N, mechanical quality factor Q_m of 800 and dielectric dissipation factor $tan\delta$ of 0.5%. Bronze is chosen for making the electrodes, and steel for the vibrator's other metal parts.

4.4.1.2 *Working principle*

Applied with a sinusoidal voltage which has a frequency close to natural frequency of the 2nd bending mode (in the *z*-directional) of the vibrators, piezoelectric plates *3* can excite a *z*-direction vibration in feet *A* and *B* in the same *z*-direction. Applied with a cosine voltage which has a frequency close to natural frequency of the 2nd bending mode (in the *y* direction), piezoelectric plates *4* can excite a *y*-direction vibration in feet *A* and *B,* in the same *y*-direction. When these two voltages are applied simultaneously, the resultant of the two linear orthogonal vibrations at the tip of each driving foot is an elliptical motion. Both the tips of feet *A* and *B* will move along an elliptical trajectory, as shown in Fig. 4.4.2. Due to the symmetry of the vibration excitation structure, they drive the vibrators or the actuator to move linearly along the guide (the guide is fixed on a solar panel).

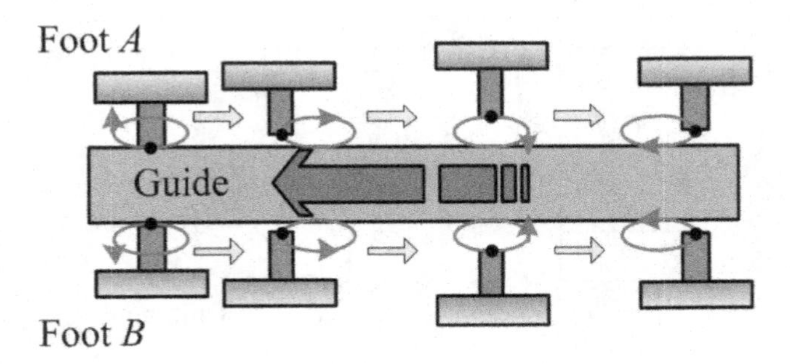

Fig. 4.4.2. Working principle of the actuator.

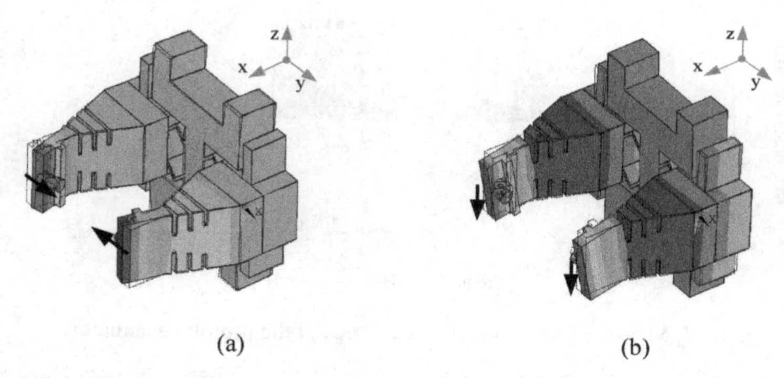

Fig. 4.4.3. Vibration s of the vibrator. (a) Vibration pattern I, and (b) Vibration pattern II.

Based on the size parameters shown in Fig. 4.4.1(b) and above listed material properties, the prototype actuator's vibration pattern is computed by the FEM with ANSYS software. Fig. 4.4.3(a) shows vibration pattern I excited by piezoelectric plates *4*. And it can be seen that the vibrators with feet *A* and *B* bend at the 2nd mode and driving feet *A* and B (two tips of the vibrators) have reverse *y*-directional movements. Fig. 4.4.3(b) shows vibration pattern II excited by piezoelectric plates *3*, and it can be seen that the vibrators with feet *A* and *B* also bend at the 2nd vibration mode but the two driving feet have the same *z*-directional movements. Calculated natural frequencies for patterns I and II are around 20.2 kHz. Results in Fig. 4.4.3 confirm the proposed actuator's working principle.

Vibration characteristics of the prototype actuator are measured by a laser Doppler vibrometer system (PSV-300F-B). For measuring vibration pattern I, measurement area *S1* (see Fig. 4.4.1(a)) is chosen and piezoelectric plates *4* are driven by a 50 $V_{0\text{-}p}$ operating voltage while the other piezoelectric plates are short-circuited. For measuring vibration pattern II, measurement area *S2* (see Fig. 4.4.1(a)) is chosen, and piezoelectric plates *4* are driven by the same conditions. Fig. 4.4.4 shows the measured results of the average vibration magnitude versus operating frequency. It can be seen that the resonance frequencies for patterns I and II are 17.4 kHz and 18.6 kHz, respectively. Measured frequency values are somewhat smaller than the calculated ones because of simplified

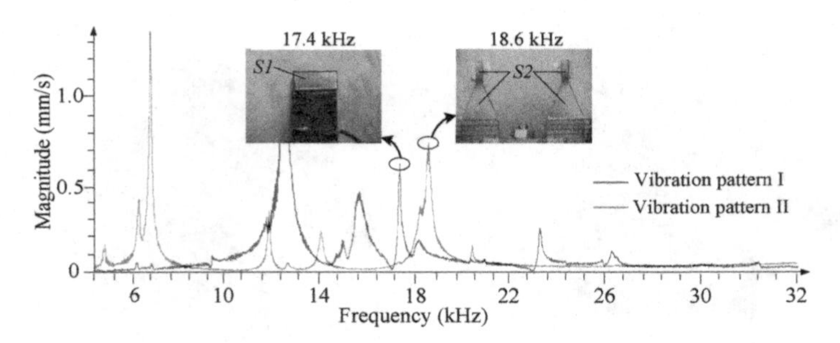

Fig. 4.4.4. Measured vibration characteristics of the prototype actuator.

model used in the calculation. At the resonance points of patterns I and II, measured vibration amplitudes are 1.2 μm and 1.6 μm, respectively.

4.4.1.3 *Assembling and mechanical characteristics of the actuator*

Based on the above design, a prototype of the linear piezoelectric actuator is fabricated and assembled onto a guide which will be fixed to a solar panel, as shown in Fig. 4.4.5(a). As shown in Fig. 4.4.5(b), the prototype actuator is fixed onto a base, and its two driving feet are pressed face-to-face onto the two sides of the guide. The two sliders bolted onto the base, are coupled with the guide and can linearly move along the guide. The two sliders stabilize the vibrators' movement along the guide. In the experiments of this work, unless otherwise specified, the preload between the driving feet and guide is 15 N, and operating

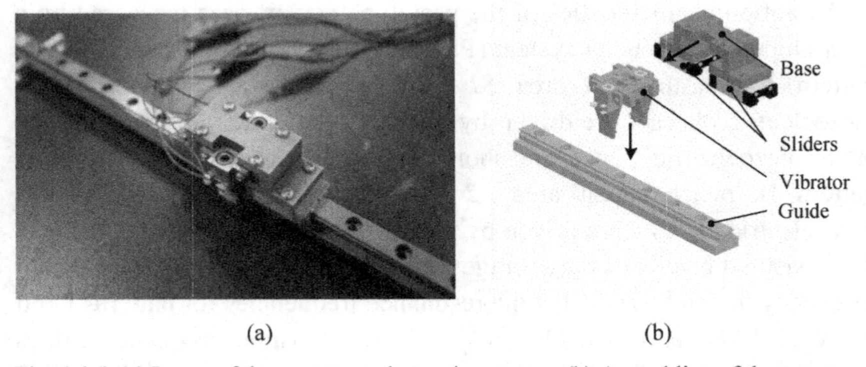

Fig. 4.4.5. (a) Image of the prototype ultrasonic actuator. (b) Assembling of the actuator.

frequency and voltage of the prototype actuator is 18.8 kHz and 100 $V_{0\text{-}p}$, respectively.

The speed-load curves for the forward and backward driving were measured at 18.8 kHz operating frequency and 100 $V_{0\text{-}p}$ operating voltage, and the result is shown in Fig. 4.4.6. For the forward driving, the no-load speed is 117 mm/s and stalling load is 3.2 N; for the backward one, they are 113 mm/s and 3.2 N, respectively. It means that the prototype piezoelectric actuator works quite well in the two opposite driving directions.

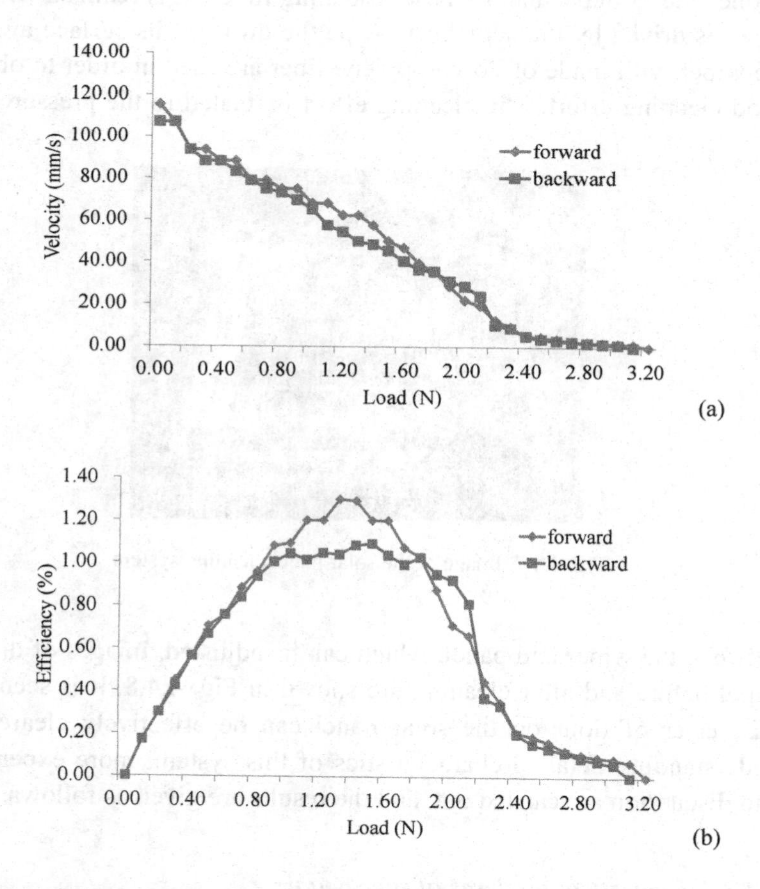

Fig. 4.4.6. Mechanical characteristics of the prototype actuator. (a) Velocity vs. load. (b) Efficiency vs. load.

4.4.2 *Solar panel cleaning system*

4.4.2.1 *Structure of the cleaning system*

Based on the prototype actuator, a solar panel cleaning system is built as shown in Fig. 4.4.7. It mainly consists of a solar panel with active area of 260 mm×300 mm (GHM-10, Changsha Guanghe Solar Co., Ltd.), villus wiper with an active length of 300 mm, and the linear actuator. The guide is fixed to the frame of the solar panel, and the wiper is assembled to one side of the actuator's base. Cleaning function is realized when the wiper is driven by the actuator to wipe the dusts on its surface away. In the wiper, villi made of Polypropylene fiber are used in order to obtain a good cleaning effort. The cleaning effort is related to the pressure force

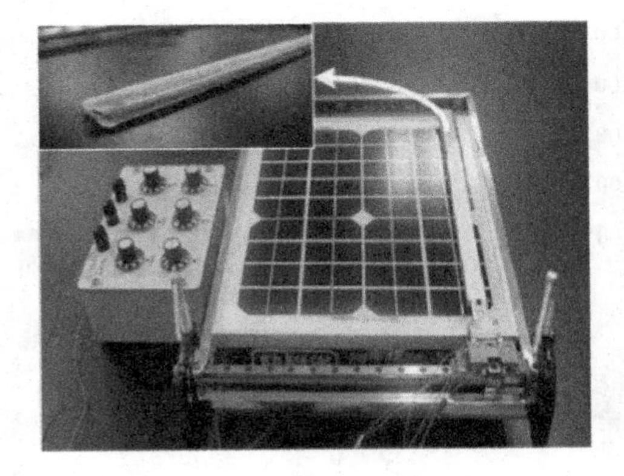

Fig. 4.4.7. Image of the solar panel cleaning system.

between the wiper and panel, which can be adjusted. Images of the solar panel before and after cleaning are shown in Fig. 4.4.8. It is seen that a dust layer of flour on the solar panel can be effectively cleared. For understanding detailed characteristics of this system, more experiments and discussion are carried out, and the results are given as follows.

4.4.2.2 *Theoretical analyses of energy gain*

Increase of electric output power of the solar panel due to the cleaning is

$$\Delta P = \frac{U_1^2}{R_1} - \frac{U_0^2}{R_0} = \Delta\Gamma \times A_s \tag{4.4.1}$$

where R_n (n = 0, 1) is the optimum load resistance for acquiring the largest output power from the solar panel, U_0 and U_1 are the voltage across the load resistance before and after the cleaning, respectively, $\Delta\Gamma$ is the increase of power per unit area and A_s is the solar panel's active area. The electrical input power of the prototype actuator is

$$P_{in} = V_{rms} \times I_{rms} \times cos\theta \tag{4.4.2}$$

where V_{rms} and I_{rms} are the rms values of the input voltage and current, respectively, and θ is the phase angle between the input voltage and current. For the actuator, there is

$$P_{out} = P_{in} \times \eta = P_a \tag{4.4.3}$$

where P_{out} is the mechanical output power, η is the power efficiency, and P_a is

$$P_a = f_c \times v \tag{4.4.4}$$

where v is the actuator's speed and f_c is the load force (the friction force between the wiper and solar panel).

Before cleaning After cleaning

Fig. 4.4.8. A solar panel before and after cleaning.

Energy gain of the solar panel cleaning system is defined as

$$\eta_s = \frac{\Delta P \times 3600}{P_{in} \times N \times \Delta t} \tag{4.4.5}$$

where N is the cleaning times per unit time (/h), and Δt is the actuator's operating time in each cleaning. When the energy gain is larger than 1, increase of output electric energy of the solar panel due to the cleaning is larger than energy consumption of the actuator. In practical applications, as working environment of a solar panel becomes dusty, the cleaning times per unit time need to be increased to keep the active area clean. By substituting (4.4.3) into (4.4.5), there is

$$\eta_s = \frac{\eta \times \Delta P \times 3600}{P_a \times N \times \Delta t} \tag{4.4.6}$$

From Fig. 4.4.6(b), it is known that the prototype actuator's power efficiency η can be fit by

$$\eta = -af_c^2 + bf_c \tag{4.4.7}$$

where a and b are positive. Substituting (4.4.4) and (4.4.6) into (4.4.7), there is

$$\eta_s = \frac{(-af_c + b) \times \Delta P \times 3600}{v \times N \times \Delta t} \tag{4.4.8}$$

The prototype actuator's stroke is $S = v \times \Delta t$, thus

$$\eta_s = \frac{(-af_c + b) \times \Delta P \times 3600}{N \times S} \tag{4.4.9}$$

Considering the solar panel's surface area is

$$A_s = S \times L \tag{4.4.10}$$

where L is the effective panel width perpendicular to the wiping direction or the effective wiper length. Hence

$$\eta_s = \frac{(-af_c + b) \times \Delta\Gamma \times L \times 3600}{N} \tag{4.4.11}$$

Therefore for a given solar panel and actuator, the energy gain is affected by the load force f_c, effective panel width perpendicular to the wiping direction, and cleaning times per unit time.

The relationship between the speed v and load force f_c of a linear ultrasonic motor can be fitted by

$$v = -c\ f_c + d \tag{4.4.12}$$

where c and d are positive. Electric input power of the linear actuator is

$$P_{in} = \frac{f_c(-cf_c + d)}{-af_c^2 + bf_c} = (1 + \frac{ad - bc}{-acf_c + bc})\frac{c}{a} \tag{4.4.13}$$

From the measured characteristics (Fig. 4.4.6) of the prototype, it is deduced that a = 0.005 N^{-2}, b = 0.016 N^{-1}, c = 0.039 m/s/N and d = 0.117m/s. From Fig. 4.4.6a and the measured speed of the wiper at a 1N-pressure force between the wiper and panel, it is known that f_c = 1.6 N. Thus the electric input power of the actuator is calculated to be 6.8 W.

4.4.2.3 *Experimental results and discussion*

In the measurement of characteristics of the system, the following experimental conditions are used. Flour is evenly distributed on the solar panel's working surface to simulate a dust layer, and its surface-density is 256 g/m^2. An incandescent lamp above the solar panel is used to produce 8420 lux light (measured by light meter TES-1339R) for the solar panel. Unless otherwise specified, the pressure force between the wiper and panel is 1 N. In this case, electric input power of the actuator is 7 W (measured). In order to get the maximum output power from the solar panel, different resistance values of external resistor are used. The optimum load resistance is 200 Ω for the clean panel, and it is 550 Ω for the panel with flour. With the experimental conditions, measured electric output power of the solar panel before and after the cleaning is 0.23 W and 1.29 W, respectively. In addition, it was observed that the electric output power increases gradually during the cleaning process.

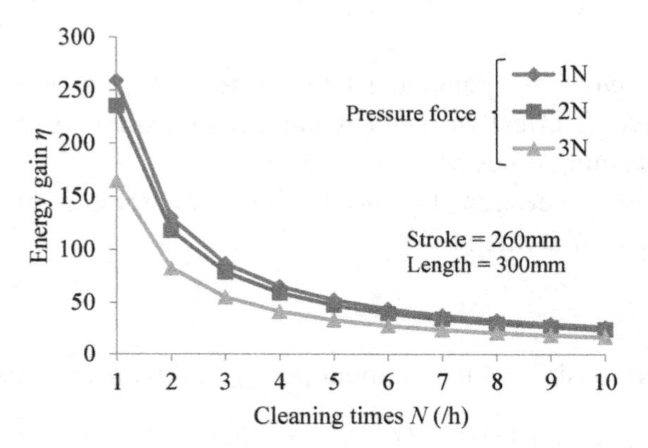

Fig. 4.4.9. Energy gain versus cleaning times per unit time at different pressure force.

Fig. 4.4.9 shows the energy gain of the solar panel cleaning system versus cleaning times per unit time N at different pressure forces between the wiper and panel. In the experiments, the highest energy gain is 252 at a pressure force of 1 N and cleaning time per hour of 1, and the lowest gain value is 16 at a pressure force of 3 N and cleaning times per hour of 10, which indicates that this cleaning system has very good energy gain.

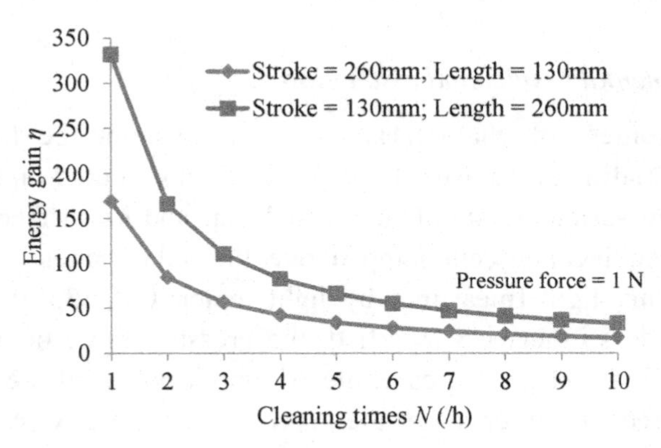

Fig. 4.4.10. Energy gain versus cleaning times per unit time at different wiper lengths.

It is observed that energy gain at a pressure force of 1N is the highest in the experiments for given cleaning times per unit time, which indicates that small pressure force is beneficial to increasing the energy gain. It is observed that the energy gain decreases with the increase of cleaning times per unit time, which means that as the environment becomes dirtier and more cleaning times per unit time are needed, the energy gain decreases. However, electric output power of the solar panel is still increased by the cleaning because the energy gain is larger than 1.

For investigating the effect of solar panel dimension L perpendicular to the wiping direction on the energy gain, a rectangular solar panel with active area of 260 mm×130 mm (GHM-5, Changsha Guanghe Solar Co., Ltd.) is employed. Measured results of the energy gain versus cleaning times per unit time for two different solar panel dimensions perpendicular to the wiping direction are plotted in Fig. 4.4.10. The energy gain at L = 260 mm is larger than that at L = 130 mm, which agrees with the prediction by Eq. 4.4.11 quite well. This means that for a solar panel with given active area, wiping the panel in the direction with a shorter wiping distance (or stroke) results in a better energy gain.

We also measured the energy gain and optimum load resistance versus dust surface-density on a rectangular solar panel identical to the one used for Fig. 4.4.10 (Stroke=260mm). From Fig. 4.4.11, the optimum load resistance increases linearly when the dust surface-density increases. For a dust surface-density of 591.72 g/m^2, the optimum load

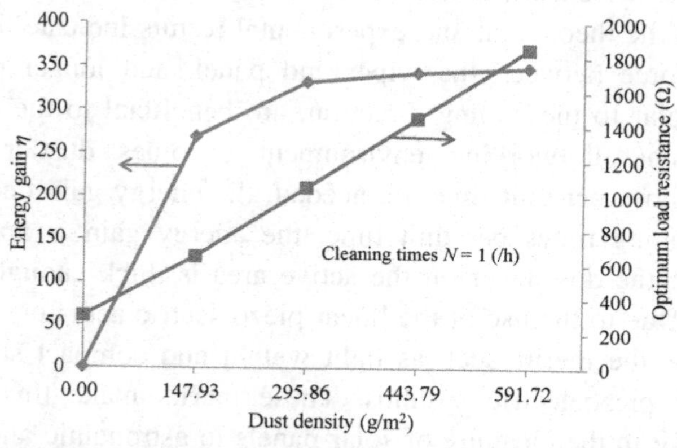

Fig. 4.4.11. Energy gain and optimum load resistance versus dust surface-density.

resistance is 1850 Ω, which is 6 times higher than that of the clean panel. It can also be seen that the energy gain first increases as the increase of the dust surface-density and then becomes stable (≈350), which is because the electric output power is approximately zero when the dust surface-density is large enough.

In practical applications, the cleaning times per hour may be set in the control system of the actuator in advance according to the environment in which the solar panel is placed. Also, an optical dust sensor may be used in the cleaning system to make the cleaning system smart. The sensor detects the dust concentration in the surroundings and sends its output signal to the control system. The control system decides the cleaning times per hour based on the dust concentration signal it receives.

4.4.3 *Summary*

A linear piezoelectric actuator based solar panel cleaning system is designed, fabricated and characterized. In the linear piezoelectric actuator, two elliptical motions of the driving feet are employed to drive the vibrator and the wiper. Excited by the driving voltage of 100 $V_{0\text{-}p}$ at resonance, the prototype actuator works well at both the forward and backward operation. The stalling load is 3.2 N and no-load speed is around 115 mm/s. In the solar panel cleaning system, the prototype actuator can drive a wiper to clear a dust layer on the solar panel's active area. Both the theoretical and experimental results indicate that a proper pressure force between the wiper and panel, and longer panel width perpendicular to the wiping direction, are beneficial to the energy gain improvement. If working environment becomes dustier and more cleaning times per unit time are needed, the energy gain decreases. For given cleaning times per unit time, the energy gain is approximately constant if the dust layer on the active area is thick enough before the cleaning. Due to the use of the linear piezoelectric actuator, the cleaning system has the merits such as light weight and compact structure, just like other piezoelectric systems. These merits make this technology competitive in the cleaning of solar panels in astronautic and aeronautic applications.

4.5 Summary and Remarks

This section demonstrates the ultrasonic methods for extraction, rotary driving and removal of micro solids, which shows that more manipulation functions can be realized by ultrasonic methods, other than the trapping. As the status quo of this technology, optimization in construction and characteristics of the devices, which is quite important for practical applications of the methods, has been seldom tried.

References

1. Hu, J. H., Yang, J. B., Xu, J., Du, J. L. (2006). Extraction of biologic particles by pumping effect in a π–shaped ultrasonic actuator, Ultrasonics, 45, pp. 15–21.
2. Hu, J. H. and Santoso, A. K. (2004). A π–shaped Ultrasonic Tweezers Concept for Manipulation of Small Particles, IEEE Trans. Ultrason. Ferroelectr. Freq. Control, 51 (11), pp. 1499–1507.
3. Barmatz, M. and Collas, P. (1985). Acoustic radiation potential on a sphere in plane, cylindrical, and spherical standing wave fields, J. Acoust. Soc. Am., 77 (3), pp. 928–945.
4. Zhou, Y. J., Li, H. Q. and Hu, J. H. (2012). An Ultrasonic Stage for Controlled Spin of Micro Particles, Rev. Sci. Instrum., 83 (4), 045004.
5. Zhang, X., Y. Zheng, Hu, J. (2008). Sound Controlled rotation of a cluster of small particles on an ultrasonically vibrating metal strip, Appl. Phys. Lett., 92 (2), 024109.
6. Zhu, X. B., Hu, J. H. (2013). Ultrasonic drive of small mechanical components on a tapered metal strip, Ultrasonics, 53 (2), pp. 417–422.
7. Lu, X. L., Zhang, Q. and Hu, J. H. (2013). A Linear Piezoelectric Actuator Based Solar Panel Cleaning System, Energy, 60, pp. 401-406.

4.5 Summary and Remarks

This section demonstrates [illegible] methods for extraction [illegible] and removal of [illegible] which shows that [illegible] [illegible] [illegible] the mapping [illegible] [illegible] is quite important for practical application of the magnetic [illegible].

References

[illegible]

[illegible]

[illegible]

[illegible]

[illegible]

[illegible]

Chapter 5

Ultrasonic Manipulations of Nanoscale Entities

This chapter demonstrates the strategies to trap, transfer and rotary drive single nanowires, and concentrate and align nanoscale entities by ultrasound. The contents of this chapter include contact type trapping and 3D transfer of a single nanowire in Section 5.1, noncontact type trapping and 2D transfer of a single nanowire in Section 5.2, controlled rotary driving of a single nanowire in Section 5.3, and concentration and alignment of nanoscale entities in Section 5.4. Acoustic streaming, which is a steady flow generated by ultrasonic field, is employed in the manipulations.

5.1 Contact Type Trapping and 3D Transfer of a Single Nanowire

This section describes a strategy for contact type trapping and 3-dimensional (3D) transfer of a single nanowire [1]. It uses mobile acoustic streaming to effectively trap and align a single nanowire within water film on substrate surface, and stably transfers a trapped nanowire through an arbitrary 3D path in the water film. The streaming is generated by a vibrating micro-probe with uniform diameter. In the experiments, a trapped nanowire is constantly on the side of the micro-probe tip, perpendicular to the micro-probe vibration and symmetric about the micro-probe approximately.

5.1.1 *Experimental setup and manipulation process*

The experiment is conducted under an optical microscopy (VHX-1000, Keyence), as shown in Fig. 5.1.1(a). In the experiments, a micro

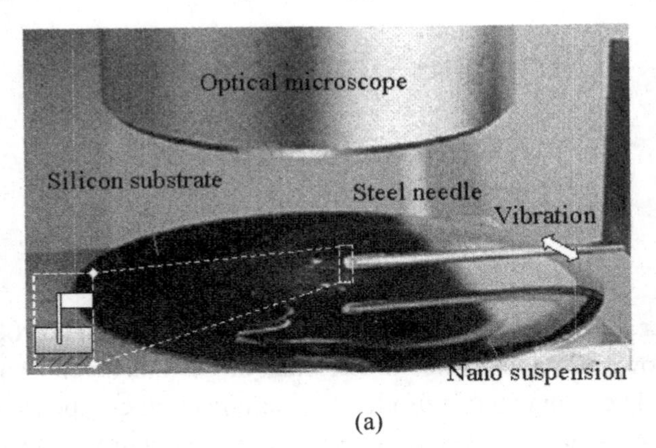

(a)

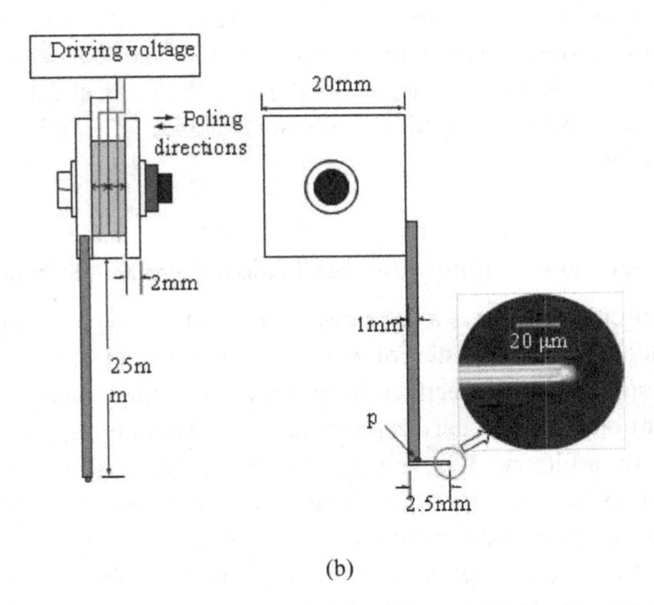

(b)

Fig.5.1.1. Experimental setup for the mobile acoustic streaming based manipulations of a single silver nanowire. (a) Schematic diagram. (b) Size and structure of the vibration excitation system.

fiberglass which is mechanically excited to vibrate by a perpendicular steel needle, is immersed into the nanowire suspension film on a silicon substrate. The fiberglass has uniform thickness of 10 μm and total length of 2.5 mm. The suspension is composed of deionized water and dispersed Silver nanowires (AgNWs) which have a diameter of 100 nm and length of several microns up to several tens of microns. The distance between the fiberglass tip and silicon substrate is 10 μ m approximately. Fig. 5.1.1(b) shows the structure and size of vibration excitation system used in our experiments. The fiberglass vibration is bonded on and excited by the tip of a steel needle which is mechanically driven by a sandwich type piezoelectric transducer. The steel needle is 25 mm long and 1 mm thick. The outer and inner diameters, and the thickness of each piezoelectric ring in the transducer are 12 mm, 6 mm, and 1.2 mm, respectively. The piezoelectric constant d_{33} is 250×10^{-12} C/N, electromechanical coupling factor k_{33} is 0.63, mechanical quality factor Q_m is 500, dielectric dissipation factor $tan\delta$ is 0.6%, and density is 7450 kg/m^3. The two stainless plates at the two ends of the transducer are square. The tightening torque applied to the transducer is 6 Nm. The operating frequency of the transducer is about 135 kHz at which the steel needle vibrates flexurally. The resonance frequency of the sandwich transducer is 93 kHz; thus the manipulation system does not operate at resonance.

Fig. 5.1.2 shows the process of driving and trapping a AgNW on the silicon substrate surface in AgNW suspension. Fig. 5.1.2(a) is a schematic diagram of the driving and trapping process, in which the origin o of the x, y and z coordinate system locates at the center of the fiberglass tip, and Fig. 5.1.2(b) is a series of snapshots to illustrate the process for a 32 μm long and 100 nm thick AgNW. The fiberglass vibrates back and forth along the y direction. The AgNW lying on the substrate surface within effect range of the vibrating fiberglass can be driven towards the point directly under the fiberglass tip (the projection point of fiberglass tip onto the substrate surface) while rotating to the fiberglass vibration direction (or the y direction), as shown in motion m1 and snapshots b1 - b4. After rotating to the fiberglass vibration direction, the AgNW continues to move to the point directly under the fiberglass tip, as shown in motion m2, and snapshots b4 and b5. Under the fiber-

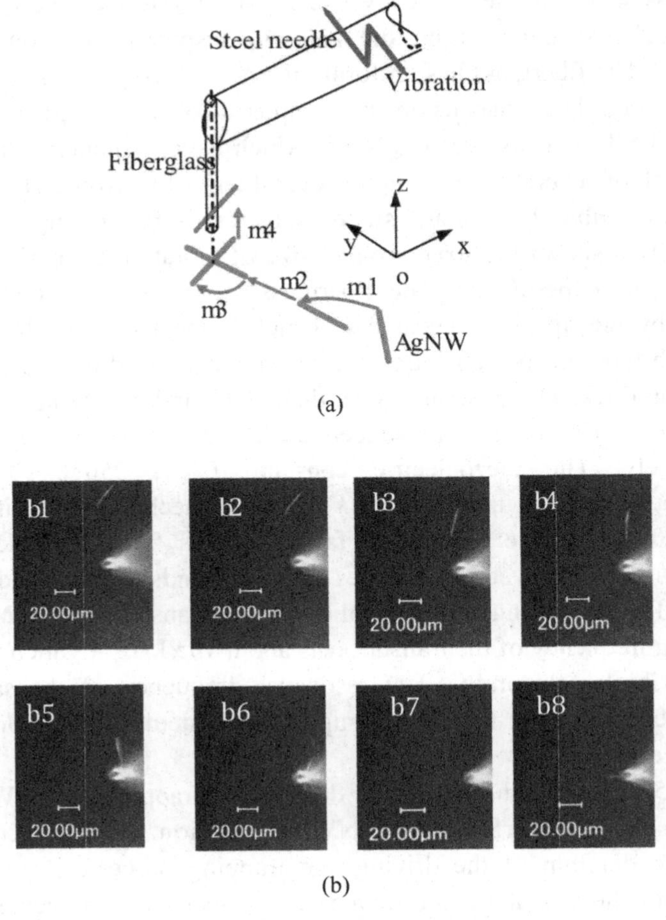

Fig. 5.1.2. Driving and trapping of a single AgNW under the fiberglass tip in water by acoustic streaming. (a) Driving and trapping of a AgNW under the fiberglass tip in water on silicon substrate surface. (b) A sequence of photographs showing different locations of the manipulated AgNW.

glass tip, the AgNW starts to rotate to the direction perpendicular to the fiberglass vibration (the x direction) while being lifted onto the side of the fiberglass tip, and eventually is pushed onto the fiberglass tip, as shown in motions m3 and m4, and snapshots b6 - b8. In snapshot b8 in Fig. 5.1.2(b), the trapped AgNW is perpendicular to the fiberglass vibration direction and approximately symmetric about the fiberglass. The approximate symmetry of trapped AgNW about the fiberglass is found by comparing the total AgNW length, the not sheltered AgNW length in b8, and the fiberglass diameter.

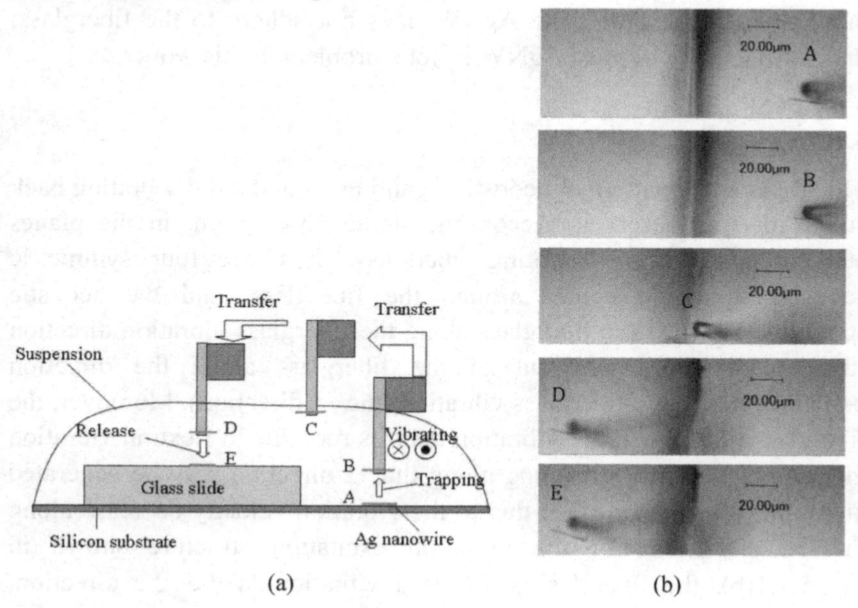

Fig. 5.1.3. Transfer and release of a single AgNW. (a) A schematic diagram. (b) A sequence of photographs showing different locations of the manipulated AgNW with length of 20 μm and diameter of 100 nm.

Fig. 5.1.3(a) is a schematic diagram to show the transfer of a AgNW with a length of 20 μm and diameter of 100 nm in the AgNW suspension, and Fig. 5.1.3(b) gives the snapshots of the AgNW at locations *A*, *B*, *C*, *D* and *E* during the transfer. The AgNW is initially at location *A* under the fiberglass tip. Then it is lifted to and trapped on the side of the vibrating fiberglass tip at location *B*. After this, the trapped AgNW is lifted and transferred to location *D* through location *C* by moving the whole manipulation device. At location *D*, the vibration is stopped and the AgNW is released to location *E* on the surface of a glass substrate. The movement of the whole manipulation device is manually conducted, using an X-Y-Z stage, and transfer speed is between several microns per second to about ten microns per second (One may use a faster transfer speed.). During the transfer, the operating frequency and voltage of the transducer is 135 kHz and 15 Vp-p, respectively. Our experiments show that the transfer can be realized along any path within the AgNW suspension. The AgNW does not adhere to the fiberglass; thus releasing the trapped AgNW is not a problem in this work.

5.1.2 *Principle*

The in-plane 2D pattern of acoustic steaming around a rod vibrating back and forth was observed. According to the observation, in the planes perpendicular to the vibrating fiberglass, there are four symmetric acoustic streaming eddies around the fiberglass, and the acoustic streaming flows to the fiberglass along the fiberglass vibration direction (the $\pm y$ direction) and out of the fiberglass along the direction perpendicular to the fiberglass vibration (the $\pm x$ direction). Moreover, the fiberglass tip has a larger vibration than its root due to flexural vibration mode; thus acoustic streaming along the $+z$ direction may be generated along the fiberglass surface due to the vibration velocity decrease along the $+z$ direction. For the vibration excitation structure shown in Fig. 5.1.1(b), the fiberglass also has a vibration in the $\pm z$ direction, which induces an acoustic streaming along the z axis for $z < 0$. Based on the above analyses, a 3D acoustic streaming pattern around the fiberglass tip is proposed, as shown in Fig. 5.1.4.

The acoustic streaming in the y direction drives the AgNW lying on the substrate surface to the location under the fiberglass tip while rotating it to the y direction. Then, under the fiberglass tip, the circumferential acoustic streaming rotates the AgNW to the x direction; meanwhile the z direction acoustic streaming lifts the AgNW to one side of the fiberglass tip. Further, the acoustic streaming in the y (or $-y$) direction pushes the lifted AgNW onto the side of the fiberglass tip. Therefore the acoustic streaming pattern shown in Fig. 5.1.4 can well explain the trapping phenomenon. The ultrasonic field around the fiberglass tip moves with the fiberglass; so the acoustic streaming field around the fiberglass tip is mobile, and can be used to trap the nanowire during the transfer.

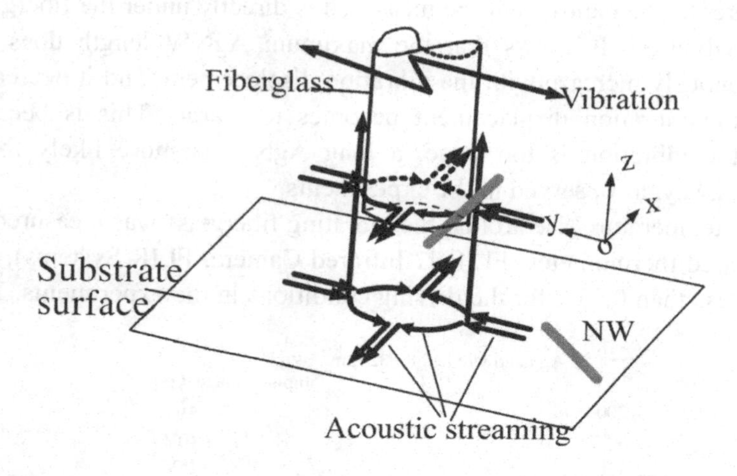

Fig. 5.1.4. The acoustic streaming field around the tip of the fiberglass in ultrasonic vibration.

5.1.3 *Characteristics and discussion*

Two parameters are defined and used for quantitative investigation of the trapping capability for the proposed method. When the distance between the initial position of a AgNW aligned in the *y* direction at $x = 0$ and the position directly under the fiberglass tip is too large, the AgNW cannot be sucked to the fiberglass tip. The maximum value of this distance d_m for sucking a AgNW is used to express the trapping capability. Another parameter to express the trapping capability is the maximum length L_m of a AgNW which can be sucked to the fiberglass tip, initially aligned in the *y* direction at $x = 0$. An AgNW 32 μm long and 100 nm thick was used in the experiments. Fig. 5.1.5(a) shows the measured dependence of maximum distance dm on the driving frequency at driving voltages of 5, 10, 15, 20, 25 and 30 Vp-p. It shows that the maximum distance reaches the maximum at the resonance point (about 135 kHz); as the driving voltage increases, the maximum distance increases. These phenomena should result from the increase of acoustic streaming speed as the fiberglass vibration increases. Fig. 5.1.5(b) shows the measured dependence of the maximum AgNW length L_m on the vibration displacement at reference point *p* [see Fig. 5.1.1(b)]. In the measurement, the center of the nanowire to be measured is directly under the fiberglass tip in vibration. It shows that the maximum AgNW length does not monotonously increase with the vibration displacement, and it decreases when the vibration displacement becomes too large. This is because when the vibration is too large, a long AgNW is more likely to be flushed away, as observed in the experiments.

The temperature rise around the vibrating fiberglass was measured by an infrared thermometer (FLIR i7 Infrared Camera, FLIR Systems), and it was less than 0.1℃ for the driving conditions in the experiments. This

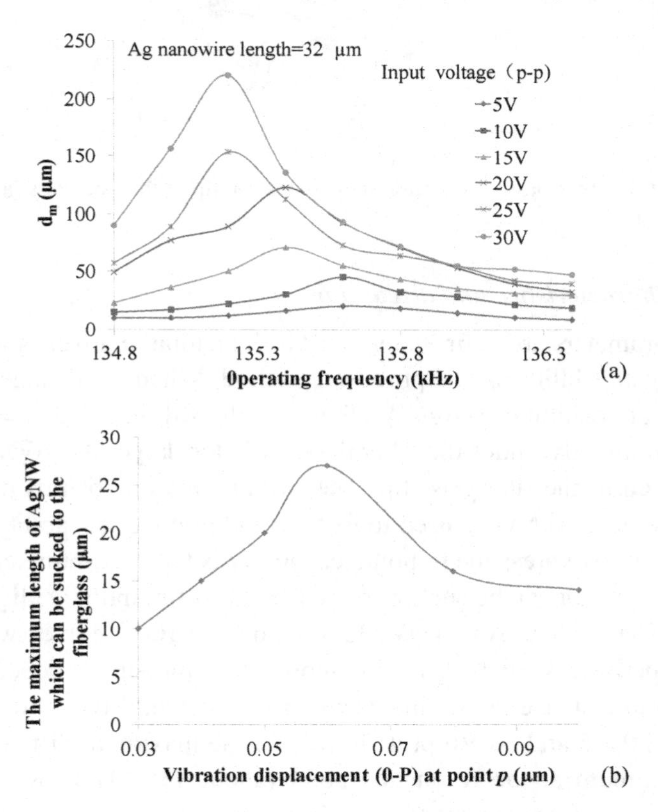

Fig. 5.1.5. The trapping capability characteristics. (a) The largest distance d_m at which a single AgNW, initially aligned in the y direction at $x = 0$, can be sucked to the fiberglass versus the operating frequency for driving voltages of 5, 10, 15, 20, 25 and 30 Vp-p. (b) The maximum length of an individual AgNW which can be sucked onto the fiberglass tip versus the vibration displacement at reference point p.

could be caused by the acoustic streaming eddies around the fiberglass tip which carry the heat away. Also, mobile acoustic streaming field around the tip of a micro-probe with other shape and vibration modes was observed. It was found that the flow pattern around the vibrating tip depends on the shape and vibration mode of the micro-probe, and more manipulation functions could be achieved by changing the acoustic streaming pattern. The details of those discoveries will be reported later, together with simulated results for the acoustic streaming around the micro-probes.

5.1.4 *Summary*

It is demonstrated that the mobile acoustic streaming is capable of efficiently driving, aligning and trapping an individual nanowire in water film on the substrate surface and stably transferring the trapped nanowire through an arbitrary 3D path in water film. In the experiments, temperature rise at the trapping position is very low (< 0.1℃), which is useful in the manipulations of thermo-sensitive samples. By controlling the mobile acoustic streaming pattern near the manipulating micro-probe, it is possible to implement more nano-manipulation functions.

5.2 Noncontact Type Trapping and 2D Transfer of a Single Nanowire

This section describes a strategy for noncontact type trapping and 2D transfer of a single nanowire [2]. It uses an ultrasonic micro-beak to suck, align, trap and transfer a single nanowire in water film on the surface of a glass substrate. The sucking and trapping force is generated by symmetric acoustic streaming eddies flowing into the micro-beak tip in ultrasonic vibration, from the front of the micro-beak and along the direction perpendicular to the micro-beak vibration. A nanowire in front of the micro-beak can be sucked to and trapped under the micro-beak, and aligned perpendicularly to the vibration direction approximately. A trapped nanowire, which is not in contact with the micro-beak, can be dragged on the substrate surface by moving the micro-beak, along an arbitrary 2D path in the water film.

5.2.1 *Experimental setup, method and phenomena*

The experiments were conducted under an optical microscopy (VHX-1000, Keyence), as shown in Fig. 5.2.1(a). The micro-beak which is immersed into AgNW suspension film on a glass slide and assembled at the end of an acoustic needle, can suck, align and trap a single AgNW when the acoustic needle vibrates properly. The acoustic needle is made

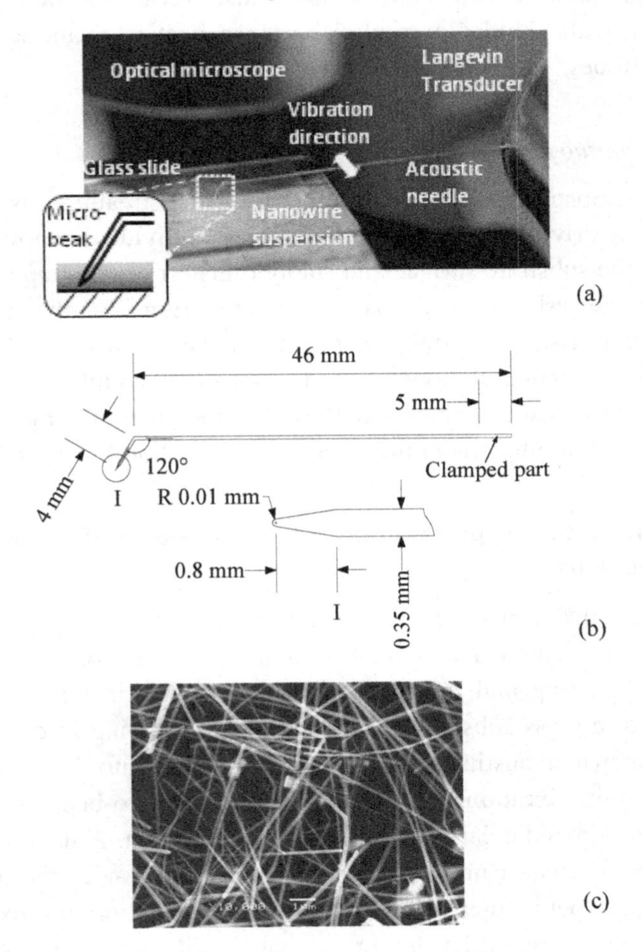

Fig. 5.2.1. Experimental setup and nano material for manipulating a single AgNW. (a) Schematic diagram of the experimental setup. (b) The shape and size of the needle with a micro-beak. (c) Image of the AgNWs used in our experiments.

of stainless steel, and its shape and size are shown in Fig. 5.2.1(b). It is mechanically excited at its root by a Langevin transducer, and has the diameter of 0.35 mm and total length of 50 mm. Vibration excitation part of the acoustic needle is clamped by the Langevin transducer and a 2 mm-thick aluminum circular plate, and the length of the clamped part is 5 mm. Bent at 4 mm from the tip, the micro-beak has an angle of 120°. The tip of the micro-beak has spherical surface with 10 μm diameter, and the tapered length is 0.8 mm. Operating frequency of the whole actuator is 110.3 kHz; resonance frequency of the Langevin transducer is 100.1 kHz; thus the manipulating system does not operate at resonance, which results in a relatively high operation stability. Fig. 5.2.1(c) shows the image of AgNWs used in the suspension film, taken by Scanning Electron Microscope (SEM). The AgNWs are 20 ~ 70 μm long with about 100 nm diameter. AgNWs are dispersed in deionized water with a concentration of 0.001 mg/ml. The suspension film thickness in the experiments is about 1.2 mm. The distance between the acoustic needle tip and glass substrate is about 15 μm. The vibration displacement distribution of the acoustic needle along its length direction was measured by laser vibrometer POLYTEC PSV-300F at the operating frequency, and the result is shown in Fig. 5.2.2. It is seen that the acoustic needle vibrates flexurally. Fig. 5.2.3 shows the structure and size of the ultrasonic transducer used in the experiment.

Fig. 5.2.4 shows the trapping process of a single AgNW 43 μm long and 100 nm thick, implemented by the vibrating micro-beak immersed in water film on the substrate surface. The operating frequency and voltage of the transducer are 110.3 kHz and 15 V, respectively. Photo *a* shows the initial state of the process, in which an AgNW in the top left-hand corner and the vibrating micro-beak tip in the bottom right-hand corner are 309 μm away. Moving the vibrating micro-beak to the AgNW, we have the state shown in photo *b*, in which the distance between the micro-beak tip and AgNW is 58 μm. At this position, the AgNW starts to rotate to the central axis direction of the acoustic needle, as shown in photos *c-e*. From photo *c* to *e*, the AgNW rotates 39.28 deg in 1 min 37 s. Then, in photos *f* and *g*, the AgNW moves to the micro-beak tip, during which it moves 40.19 μm in 2 min. In photo *h*, the AgNW is under the micro-beak tip, and they are not in contact. In photos *h-k*, the trapped

AgNW is dragged to the bottom right-hand corner through a 2D path on the substrate surface by moving the micro-beak and transducer. In 10 min 13 s, the trapped nanowire is dragged by a linear distance of 277 μm. Photo *l* shows the release of the AgNW after switching the vibration off.

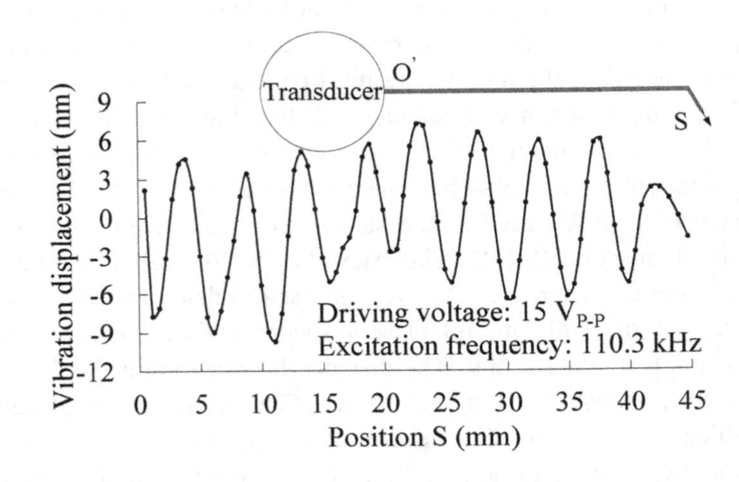

Fig. 5.2.2. Vibration displacement distribution along the needle length from its root.

5.2.2 *Operation principle*

Using deionized water suspension of 500~800 nm thick SiC fibers and their agglomerations, the acoustic streaming around the micro-beak was observed, and the observed flow pattern is shown in Fig. 5.2.5. In the figure, the origin of the xyz coordinate system is at the center of the micro-beak tip. It is seen that the acoustic streaming pattern is symmetric about the x axis; on the substrate surface, the acoustic streaming flows to the micro-beak tip from all directions; it flows upwards along the micro-beak surface and then outwards. Also, the acoustic streaming speed on the substrate surface is not uniform for different directions, and it is the fastest in the micro-beak vibration direction (the $\pm y$ direction), and the weakest in the $-x$ direction.

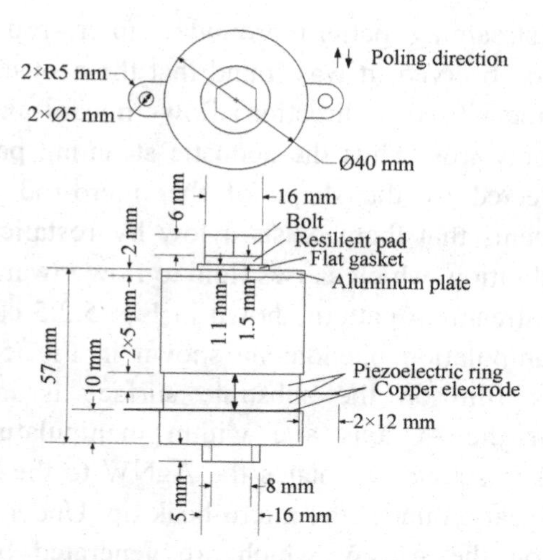

Fig. 5.2.3. The structure and size of the ultrasonic transducer.

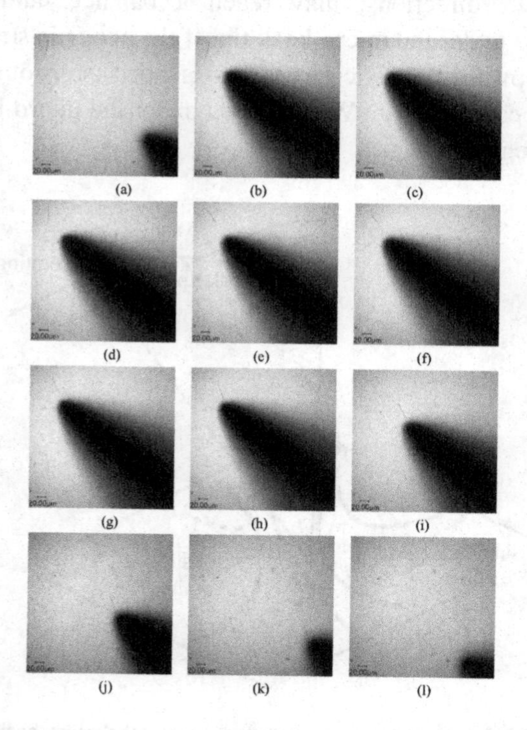

Fig. 5.2.4. Manipulation process of a single AgNW in water film on the surface of a glass substrate.

The acoustic streaming pattern around a micro-rod with uniform diameter was also observed. It was found that the acoustic streaming in the y direction (the vibration direction) flows inward and that in the x direction flows outwards. Thus the acoustic steaming pattern around a micro-rod is affected by the shape of the micro-rod. For a conical micro-beak, it seems that there exists a low hydrostatic pressure zone under its tip in vibration, which causes fluid to flow inward.

The acoustic streaming pattern shown in Fig. 5.2.5 can well explain the observed manipulation phenomena shown in Fig. 5.2.4. When an AgNW in water film on the substrate surface is in front of the micro-beak, near the $-x$ axis and within manipulating region, the x-direction acoustic streaming rotates the AgNW to the x direction and pushes it to the location under the micro-beak tip. Under the micro-beak tip, the forces on the AgNW which are generated by the acoustic streaming in $\pm x$ directions, may reach a balance, and the AgNW eventually stops under the micro-beak tip. If the acoustic streaming is not too fast, the upward acoustic streaming is not fast enough to lift the AgNW. In this case, the AgNW is trapped under the micro-beak (without contact with the micro-beak).

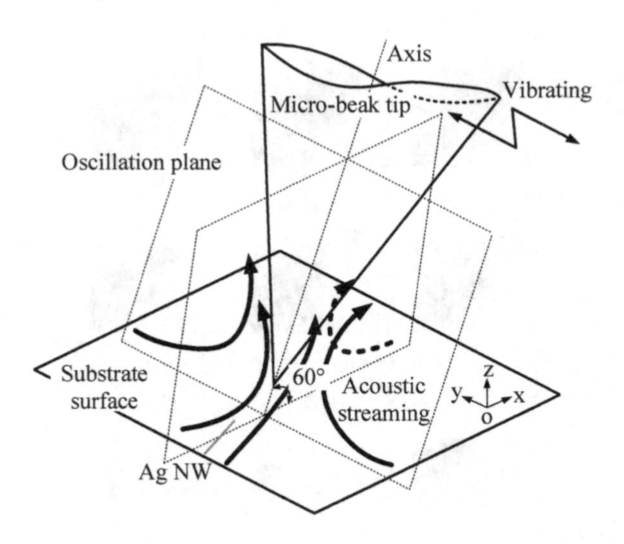

Fig. 5.2.5. Acoustic streaming pattern around the micro-beak tip.

5.2.3 *Characteristics and discussion*

For an AgNW on the substrate surface in water film, which is in front of the micro-beak and near the $-x$ axis, the acoustic streaming can push the AgNW to the micro-beak if it is not too far away from the micro-beak tip. Defining the maximum distance between a AgNW and the point directly under the micro-beak tip for sucking the AgNW as the maximum trapping distance, its dependence on the vibration displacement of reference point *o'* (see Fig. 5.2.2) in the direction shown in Fig. 5.2.1(a) was measured for three AgNWs with identical thickness of 100 nm and different lengths, and the result is shown in Fig. 5.2.6. It shows that the maximum distance increases as the vibration increases, which can be explained by the speed increase of acoustic streaming in the x direction. In the same experiment, it was observed that when the vibration displacement is too large, nanowires in front of the micro-beak are usually flushed away, and the trapping is not stable. In our experiments, the maximum trapping distance is about 150 μm or so. In practical applications, this distance may greatly reduce the precision requirement for driving motors in a manipulation system.

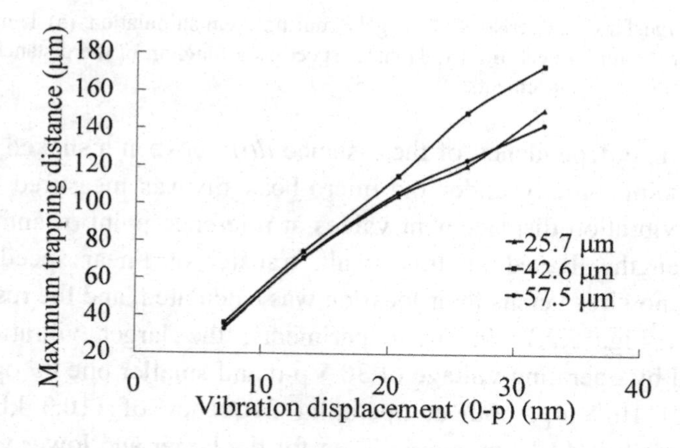

Fig. 5.2.6. The maximum trapping distance versus vibration displacement amplitude at reference point *o'* for single AgNWs with different length.

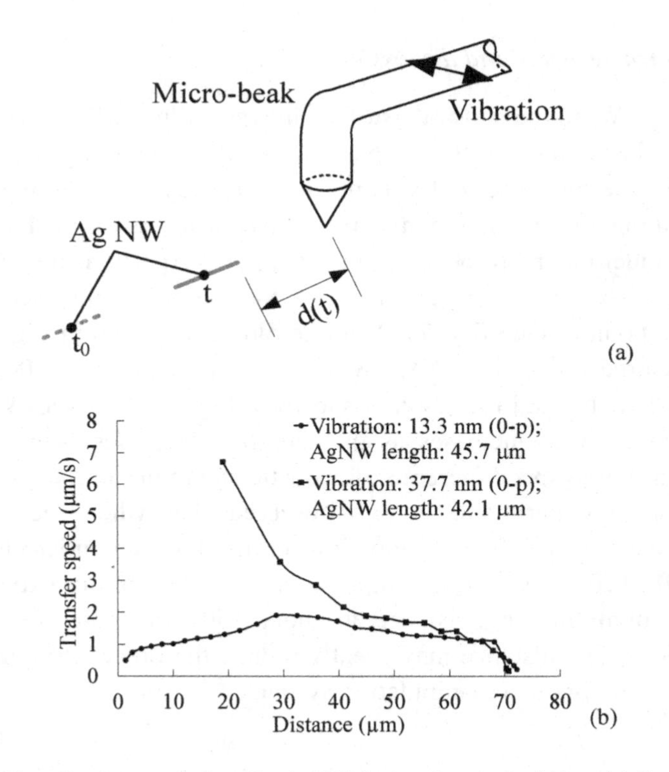

Fig. 5.2.7 Transfer characteristics of a AgNW during the manipulation. (a) Transfer of a AgNW near the micro-beak tip. (b) Transfer speed as a function of the distance for two different vibration displacements.

The time dependency of the distance *d(t)* between a sucked AgNW and the point directly under the micro-beak tip was measured for two different vibration displacement values at reference point *o'* and similar AgNW lengths. Based on this result, transfer or linear speed of the sucked nanowires versus their location was calculated, and the results are shown in Fig. 5.2.7. In the experiments, the larger vibration was generated by operating voltage of 30 Vp-p and smaller one by operating voltage of 10 Vp-p, both at operating frequency of 110.3 kHz. The AgNW length is 42.1 μm and 45.7 μm for the larger and lower vibration operation, respectively. For the smaller vibration operation, as the AgNW approaches the micro-beak tip, its speed increases first and then decreases to zero, which means that the sucked nanowire stops under the micro-beak tip. This is because the forces induced by $\pm x$ direction acoustic streaming can reach a balance at smaller vibration. For the larger vibration operation, as the AgNW approaches the micro-beak, its speed increases monotonously, which means that the nanowire cannot stop under the micro-beak tip. In this case, the AgNW will be flushed away. This is because the $\pm x$ direction acoustic streaming has a large difference in speed at larger vibration, and the forces on the nanowire which are induced by the flows, cannot reach a balance.

Based on the experimental results given in Fig. 5.2.7(b), the driving force on the nanowire may be estimated. For the AgNW moving to the micro-beak tip, as shown in Fig. 5.2.7(a), the driving force F_d is

$$F_d = F_f + ma \tag{5.2.1}$$

where F_f is the friction force on it, m is its mass and a is its acceleration. Denoting the friction coefficient between a AgNW in water and glass slide substrate as μ, the friction force is

$$F_f = \mu\,(mg - V\rho_w g) \tag{5.2.2}$$

where V is the volume of the AgNW, ρ_w is the water density (=1000 kg/m^3), and g is the gravitational acceleration (= 9.8 m/s^2). Tilting the glass slide substrate and measuring the tilting angle θ_c at which the AgNWs in water film moves very slow on the glass slide substrate, the friction coefficient between an AgNW in water and glass slide substrate can be calculated by $\mu = \mathrm{tg}\theta_c$. Based on our measurement, it is 0.64 approximately. For the AgNW with a length L of 42.1 μm and diameter d of 100 nm, volume V is 3.3×10^{-19} m^3 (= $\pi L d^2/4$), and mass m is 3.47×10^{-15} kg (the density of silver is 10.5×10^3 kg/m^3.); for the AgNW with the length L of 45.7 μm and diameter d of 100 nm, volume V is 3.6×10^{-19} m^3, and mass m is 3.78×10^{-15} kg. From Fig. 5.2.7(b), the acceleration versus time can be calculated, and the result is shown in Fig. 5.2.8. Based on the above analyses, $F_f/(ma)$ is calculated to be 1.4×10^8 at t = 2 s for the 42.1 μm-long nanowire at vibration displacement of 37.7 nm (0-p), and 9.3×10^7 for the 45.7 μm-long nanowire at vibration displacement of 13.3 nm (0-p), which indicates that the term

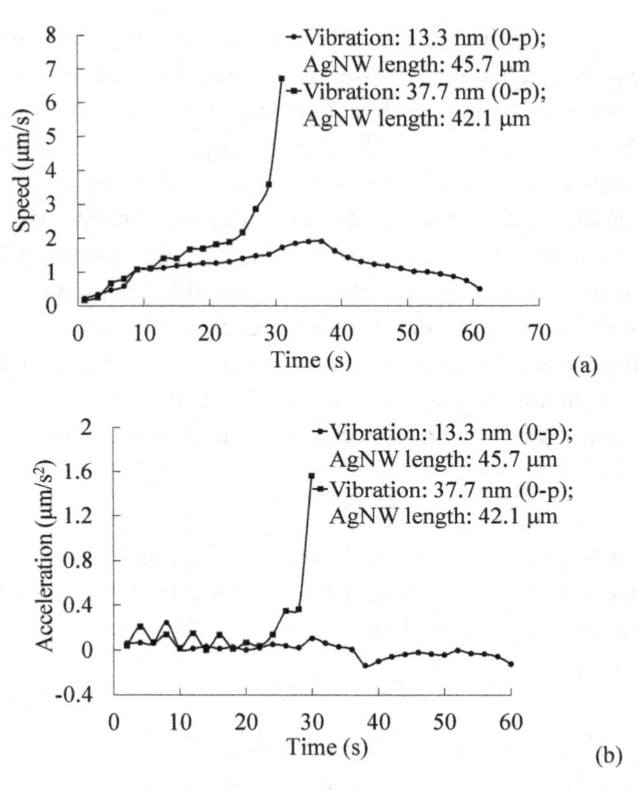

Fig. 5.2.8. Transfer speed and acceleration as a function of time for two different vibration displacements. (a) Speed. (b) Acceleration.

ma may be ignored in the estimation of driving force F_d. Also the friction force is calculated to be 0.02 pN for the 42.1 μm-long nanowire at vibration displacement 37.7 nm (0-p), and 0.022 pN for the 45.7-μm long nanowire at vibration displacement 13.3 nm (0-p). These results show that the driving force has the order of magnitude of 10^{-14} N for the nanowires in the experiments, and the order of magnitude of the driving force is less sensitive to the vibration. The latter seems to indicate that the nanowires are within the boundary layer of acoustic streaming field.

Defining the angle made by an AgNW and the x axis as orientation angle θ_o of the AgNW, the orientation angle versus time was measured for a rotating AgNW, which had the length of 67 μm, initial orientation angle of 72.5°, and initial center position of $x = -41$ μm, $y = 0$. The result is shown in Fig. 5.2.9(b). In the experiment, the vibration displacement was 13.6 nm (0-p) at reference point *o'*, and the nanowire transferred very little during the experiment. The angular speed was calculated from the curve of the orientation angle versus time, and the result is shown in Fig. 5.2.9(c). Fig. 5.2.9(c) not only shows that the angular speed is not constant during the rotation, but also indicates some interesting change of angular acceleration a_θ (slope of the curve in Fig. 5.2.9(c)) as the time increases or the orientation angle θ_o decreases. From the slope change of the curve in Fig. 5.2.9(c), it is known that as the orientation angle θ_o becomes less than about 55°, there is remarkable increase in the angular acceleration a_θ; as the orientation angle θ_o decreases further and becomes less than 35°, the angular acceleration a_θ becomes negative.

Based on these phenomena, the driving and resistance torques on the rotating nanowire are analyzed as follows. Dividing a whole nanowire with length L into two symmetric sections, the driving torque T_d is caused by difference of the flow induced torques on these two sections, and can be derived.

$$T_d = \frac{1}{8}\ \alpha L^2 sin\theta_o \tag{5.2.3}$$

where α is a parameter related to the asymmetry and largeness of acoustic streaming speed on the two symmetric sections of nanowire, and to the nanowire diameter. The angular acceleration a_θ satisfies the following equation:

$$T_d - T_r = I\ a_\theta \tag{5.2.4}$$

where T_r is the resistance torque on the nanowire, and I the moment of inertia about the nanowire center. They can be calculated by

$$T_r = 2\int_{\mathrm{r}=0}^{\mathrm{L}/2} \pi g\mu R^2(\rho_n - \rho_w) r dr = \pi g\mu R^2(\rho_n - \rho_w)L^2/4 \tag{5.2.5}$$

$$I = \frac{L^2 m}{12} \tag{5.2.6}$$

where R is the radius of the nanowire, and ρ_n and ρ_w are the density of the nanowire and water, respectively.

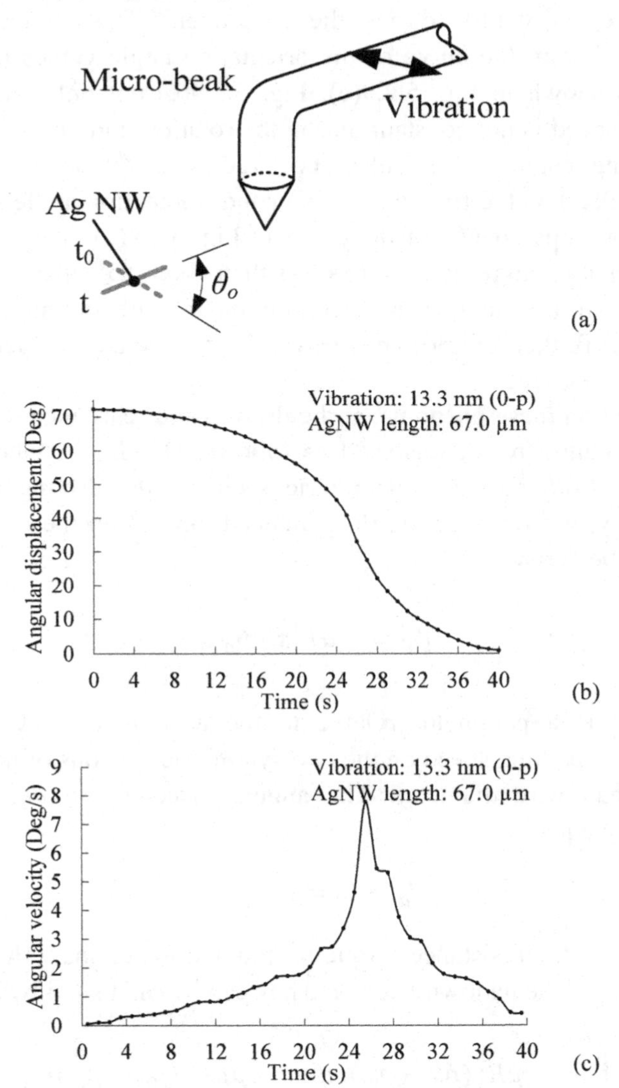

Fig. 5.2.9. Rotation characteristics of a AgNW during the manipulation. (a) Rotation of a AgNW near the micro-beak tip. (b) Angular displacement of the AgNW from the acoustic streaming direction as a function of time. (c) Angular speed as a function of time.

From Eqs. (5.2.3)-(5.2.6), it is derived that the remarkable increase of angular acceleration as orientation angle θ_o becomes less than 55° must be caused by the increase of driving torque T_d, considering the resistance torque T_r does not change with the angular speed (see Eq. (5.2.5)). The increase of driving torque T_d could be caused by an increase of parameter α, resulting from an increase of acoustic streaming speed at the nanowire as orientation angle θ_o decreases. The negative angular acceleration should result from the decrease of the driving torque T_d as orientation angle θ_o decreases (see Eq. (5.2.3)).

In addition, from the angular speed-time curve near the starting point in Fig. 5.2.9(c), one can calculate the net driving torque Ia_θ on the nanowire by Eqs. (5.2.4) and (5.2.6). In the calculation, the density of silver nanowire ρ_n is 10490 kg/m^3, the length L and radius R of the AgNW are 67 μm and 50 nm, respectively, and the angular acceleration of the nanowire at the starting is 0.00145 rad/s^2. The calculated net driving torque Ia_θ is 3×10^{-27} Nm when the vibration displacement at reference point o' is 13.3 nm (0-p). Furthermore, the resistance torque T_r may be estimated by Eq. (5.2.5), and it is 5.4×10^{-19} Nm, which is much larger than Ia_θ. Therefore, the driving torque T_d has the order of magnitude of 10^{-19} Nm.

5.2.4 *Summary*

An ultrasonic micro-beak can be used to suck, align and trap a single silver nanowire in water film on a substrate surface in a noncontact way. The sucking, aligning and trapping force is generated by the acoustic streaming eddies flowing into the micro-beak tip in ultrasonic vibration. A trapped nanowire is always under the ultrasonic micro-beak tip, perpendicular to the vibration direction of the micro-beak and aligned along the central axis direction of the acoustic needle approximately. Experimental measurement shows that the trapping range becomes larger as the micro-beak's vibration increases, and the linear and angular speeds of a sucked nanowire depend on its position relative to the micro-beak tip. The trapped nanowire, which is not in contact with the micro-beak, can be dragged to a desired location on substrate surface along an arbitrary 2D path in water film, by moving the micro-beak. The driving force on a manipulated nanowire is estimated to have an order of magnitude of 10^{-14} N. The method may have applications in the trapping and 2D transfer of individual sticky nanowires in water.

5.3 Controlled Rotary Driving of a Single Nanowire

A strategy to ultrasonically rotate a single nanowire in water film on a substrate surface is demonstrated in this section [3]. Controlled mobile acoustic streaming field, generated by a vibrating fiberglass rod, is used to move and rotate a nanowire on the substrate surface. By means of the acoustic streaming field, the nanowire center or end moves to the location directly under the tip of the fiberglass rod, and serves as the rotation center. For a 67 μm-long and 100 nm-thick silver nanowire, the revolution speed can be greater than 3 rad/s with a temperature rise less than 0.1℃ at the rotation area. The method has potential applications in the orientation and dynamics property measurement of individual nanowires, and assembling of micro/nano structures.

5.3.1 *Experimental setup and phenomena*

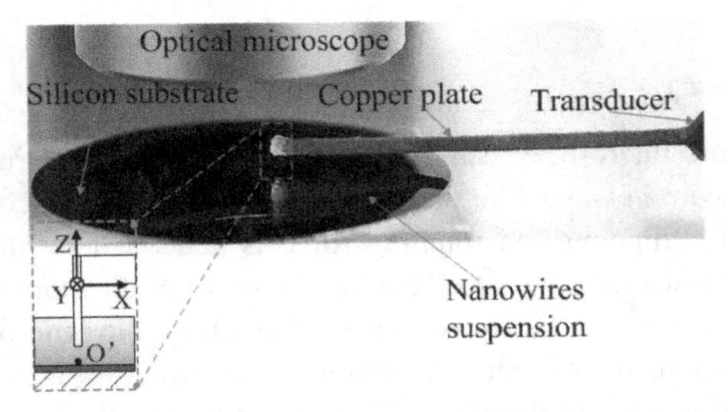

Fig. 5.3.1. Experimental setup for rotary driving of a single AgNW in water film on a silicon substrate.

The experiments are conducted under an optical microscopy (VHX-1000, Keyence), as shown in Fig. 5.3.1. In the experiments, a micro fiberglass rod which is mechanically excited to vibrate by a perpendicular copper plate is immersed into nanowire suspension film on a silicon substrate. The fiberglass rod has uniform diameter of 10 μm and total length of 3.2 mm. The suspension is composed of deionized water and dispersed AgNWs, which have diameter of about 100 nm and length of several micrometers to several ten micrometers. The distance between the fiberglass rod tip and silicon substrate is about 10 μm.

Fig. 5.3.2(a) shows the structure and size of the vibration excitation system. The fiberglass rod for generating the acoustic streaming is perpendicularly bonded on the end of a copper plate, which is mechanically driven by a sandwich type piezoelectric transducer. The outer, inner and thickness of each piezoelectric ring in the transducer are 12 mm, 6 mm and 1.2 mm, respectively. The piezoelectric constant d_{33} is 250×10^{-12} C/N, electromechanical coupling factor k_{33} is 0.63, mechanical quality factor Q_m is 500, dielectric dissipation factor $tan\delta$ is 0.6%, and density is 7450 kg/m^3. The two stainless steel plates at two ends of the transducer are square. The copper plate is bonded onto the surface of one of the square plates along the diagonal direction by adhesive material. The tightening torque applied to the transducer is 6 Nm. The operating frequency of the transducer is around 137 kHz. As the resonance frequency of the transducer is about 93 kHz, the transducer does not operate at resonance, which makes the rotary drive more stable. The image of the ultrasonic device used in our experiment is shown in Fig. 5.3.2(b).

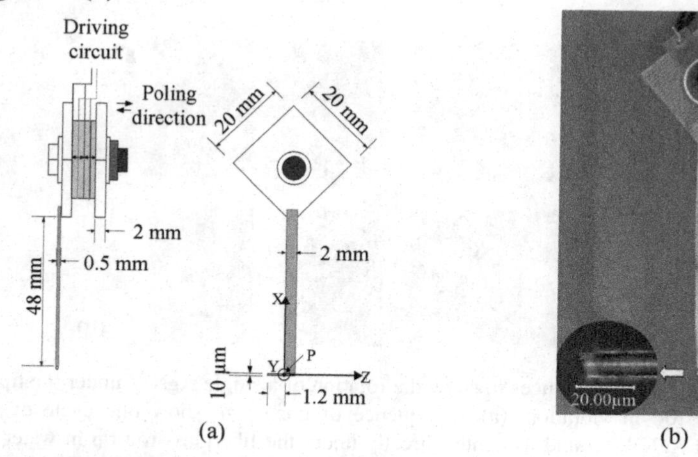

Fig. 5.3.2. Ultrasonic vibration excitation method. (a) Ultrasonic vibration excitation structure and its size. (b) Image of the ultrasonic device.

Fig. 5.3.3 shows two types of rotary driving of a single AgNW under the fiberglass rod tip in AgNW suspension on the silicon substrate, which were achieved in our experiments. In the experiment, the operating frequency and voltage are 137 kHz and 6 V_{p-p}, respectively. In Fig. 5.3.3(a), images *a*1 - *a*8 show the clockwise direction rotation (viewed from the top) of a 80 μm-long and 100 nm-thick AgNW in one cycle (14s). The AgNW center is directly under the fiberglass rod tip (see Fig.5.3.1), and the rotation is around the AgNW center. Fig. 5.3.3(b) shows another type rotation, in which a 30 μm-long and 100 nm-thick AgNW on the silicon substrate in AgNW suspension rotates around one end of itself in the clockwise direction. The rotation center or the nanowire end is directly under the vibrating fiberglass rod tip (see Fig.5.3.1). Images *b*1- *b*8 show the rotation in one cycle (8.3 s). It was observed that when an AgNW on the substrate surface was in the vicinity

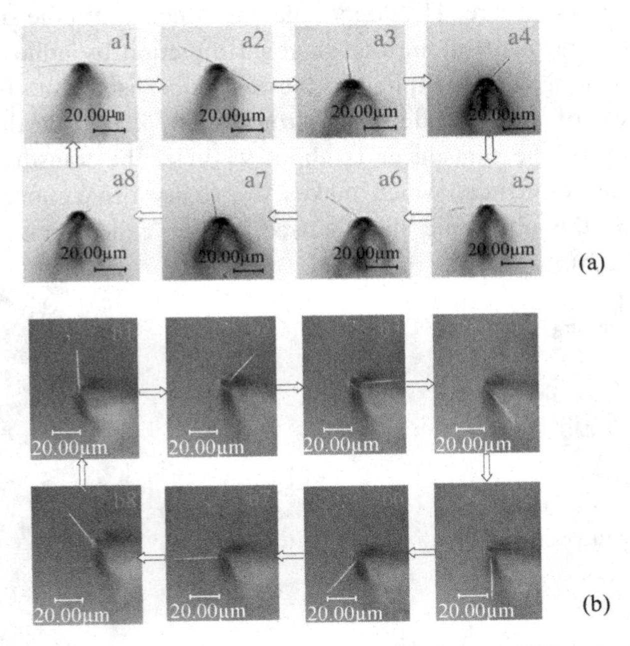

Fig. 5.3.3. Two Image sequences to show the rotation of a single AgNW under the tip of the fiberglass rod in vibration. (a) A sequence of images to show one cycle of the rotation of an AgNW around its center directly under the fiberglass rod tip in water on the substrate. (b) A sequence of images to show one cycle of the rotation of an AgNW around one end of itself directly under the fiberglass rod tip in water on the substrate.

of the fiberglass rod tip, it could shift to the position where its center or one end was directly under the fiberglass rod tip (see Fig. 5.3.1), while rotating. Hence there are two possible rotation centers or pivots for a nanowire on the substrate surface. One is its center and another is its end. Either of them is directly under the fiberglass rod tip. Also, based on many times of observation, it is concluded that whether a nanowire rotates around its center or end depends on the initial relative position between the nanowire and fiberglass rod tip. If the nanowire center is closer to the fiberglass rod tip than the nanowire end initially, the nanowire center will shift to the position directly under the fiberglass rod tip (point *O'*), and serves as the rotation center. Otherwise, the nanowire end will shift to point *O'*, and serves as the rotation center.

5.3.2 *Principle*

To under the physical principle of the above phenomena, the pattern of acoustic streaming around the fiberglass rod was experimentally investigated. In the experiment, water suspension with SiC micro-clusters was used, and the fiberglass rod in vibration was vertically inserted into the suspension film. It was observed that the SiC micro-cluster rotated around the fiberglass rod tip while moving to the point directly under the fiberglass rod tip, as shown in Fig. 5.3.4(a). This phenomenon indicates that there exist two acoustic streaming components around the fiberglass rod tip. One is in the circumferential direction around point *O'*, and another in the radial direction (pointing to point *O'*). Further observation shows that there is acoustic streaming component in the *Z*- direction along the fiberglass rod.

To understand the reason for generating the circumferential acoustic streaming, the amplitude and phase angle of the *X*- and *Y*-direction vibration displacements of point *P* at the fiberglass rod root were measured by laser Doppler vibrometer (POLYTEC PSV-300F). In the measurement, the operating frequency and voltage were 137 kHz and 10 $V_{p\text{-}p}$, respectively. The measured *X*- and *Y*-direction displacements are $U_x = 0.042sin(2\pi ft+2.25)$ μm and $U_y = 0.097sin(2\pi ft+0.69)$ μm; thus the phase difference between the two vibrations is 89.4° [=360°×(2.25-0.69)/(2π)], and the motional locus of point *P* is

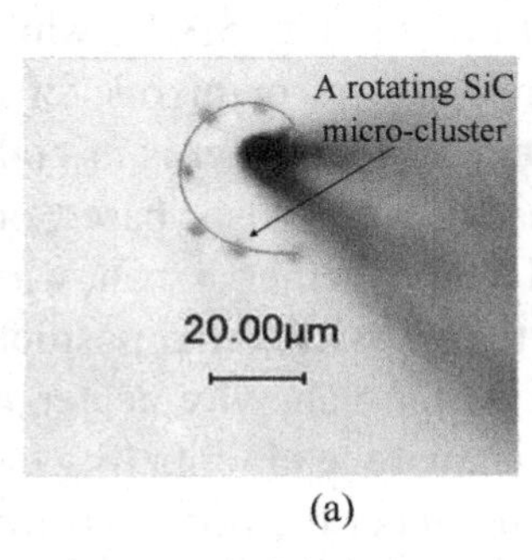

(a)

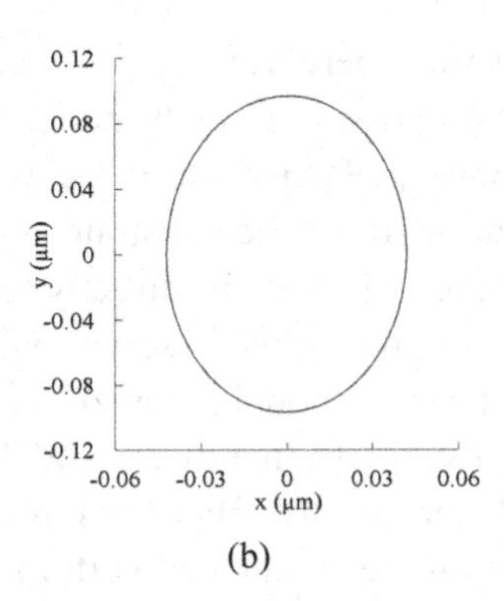

(b)

Fig. 5.3.4. Principle. (a) An image to show the acoustic streaming line around the vibrating fiberglass rod tip with a SiC micro-cluster. (b) The motional locus of the fiberglass rod root in ultrasonic vibration. The image in (a) is the overlap of 7 images taken consecutively.

approximately an ellipse parallel to the *X-Y* plane, as shown in Fig. 5.3.4(b). This motion stirs the water around the fiberglass rod and causes the circumferential acoustic streaming. The radial acoustic streaming should be caused by a low pressure area under the tip of the fiberglass rod, which is generated by the tip vibration, and the *Z*-direction one by the continuity of flow.

Based on the above experimental results and analyses, it is known that the AgNW rotation on the substrate surface under the fiberglass rod tip is caused by the circumferential acoustic streaming, and the shift of the nanowire center or end (with the whole nanowire) to point *O'* by the radial acoustic streaming.

5.3.3 *Characteristics and discussion*

The angular displacement versus driving time for an AgNW 30 μm long and 100 nm thick, rotating around the center of itself in the clockwise direction viewed from the top, was measured when the *Y*-direction vibration displacement at point *P* is 0.02 μm (0-p), and the result is shown in Fig. 5.3.5(a). In the experiment, the operating frequency and voltage are 137 kHz and 2.5 $V_{p\text{-}p}$, respectively. Fig. 5.3.5(b) shows the angular speed versus driving time, calculated from Fig. 5.3.5(a). It is seen that the angular speed is almost constant after the starting process.

During the starting process, the angular acceleration is about 0.017 rad/s^2, which is calculated from Fig. 5.3.5(b).

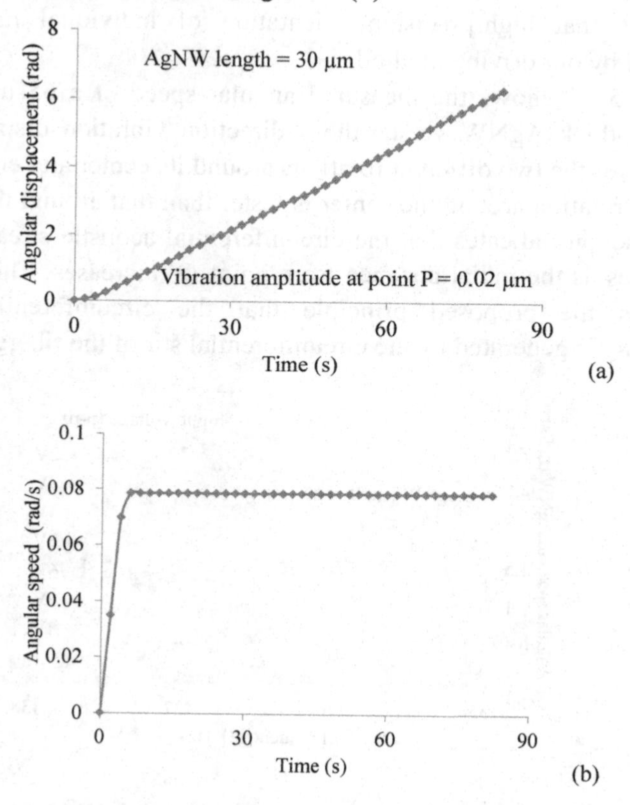

Fig. 5.3.5. The angular displacement and speed of an AgNW rotating around the center of itself versus driving time. (a) Angular displacement. (b) Angular speed.

The angular speed of an AgNW 67 μm long and 100 nm thick versus operating frequency was measured for different operating voltages, and the results are shown in Fig. 5.3.6(a). In the experiment, the nanowire rotates around its center. It is seen that the measured angular speed can be larger than 3.0 rad/s. Fig. 5.3.6(b) shows the measured angular speed of the nanowire versus the *Y*-direction vibration displacement at point *P*. It is seen that the angular speed increases with the vibration displacement. As the vibration displacement is determined by the operating frequency and voltage, the angular speed can be controlled by the operating frequency and voltage. The lowest steady angular speed obtained in our experiment is 0.001rps (see Fig. 5.3.6(a)), which indicates that high-precision orientation of individual nanowires is possible by our driving method.

Fig. 5.3.7 shows the measured angular speed of a 65 μm-long and 100 nm-thick AgNW versus the *Y*-direction vibration displacement at point *P* for the two different rotations around its center and end. It is seen that the rotation around the center is faster than that around the end. This phenomenon indicates that the circumferential acoustic streaming speed decreases as the radial distance from point *O'* increases. This deduction supports the proposed principle that the circumferential acoustic streaming is generated by the circumferential stir of the fiberglass rod.

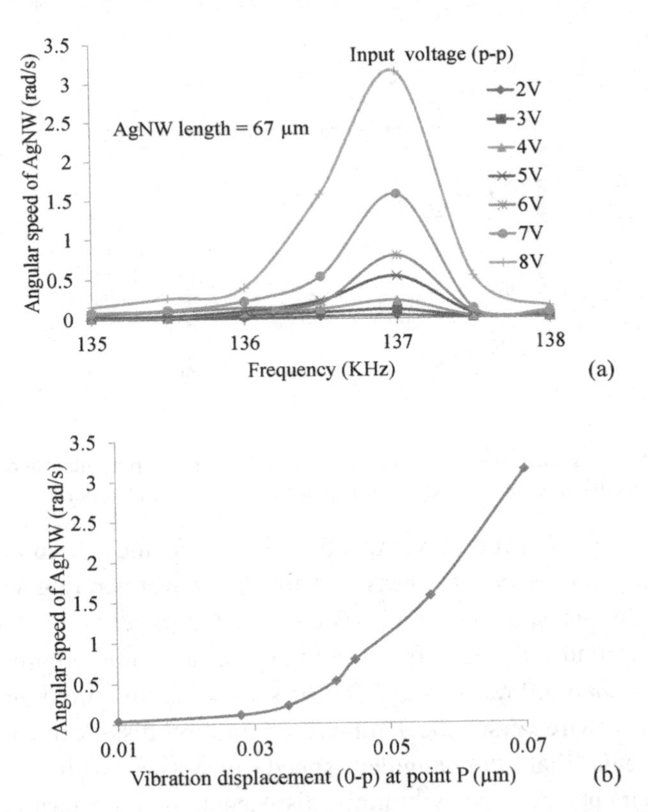

Fig. 5.3.6. The angular speed of a nanowire rotating around the center of itself. (a) For different operating frequency and voltage. (b) For different vibration displacement.

The temperature rise around the vibrating fiberglass rod was measured by infrared thermal meter (Hioki 3419-20), and it is less than 0.1℃ for the driving conditions in our experiments, which means that the method has little heat damage to biological samples. This may be caused by the acoustic streaming eddies around the fiberglass rod, which carries the heat away.

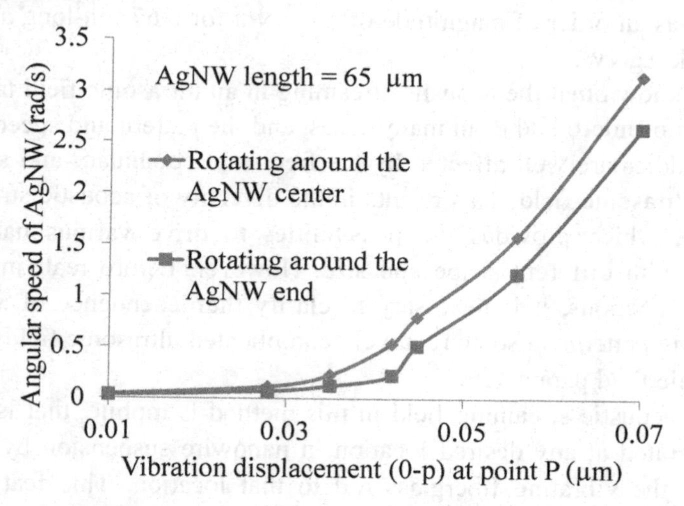

Fig. 5.3.7. A comparison of the angular speed of a nanowire rotating around its center and end.

For a nanowire rotating around its center in water suspension on the silicon substrate in our experiment, the driving and frictional torques T_d and T_f on it satisfy the following equation: $T_d = T_f + I\,\alpha_\theta$ (5.3.1), where I is the moment of inertia [= $L^2M/12$ (5.3.2), here L is the nanowire length and M the nanowire mass.], and α_θ is the angular acceleration. The frictional torque can be calculated by $T_f = \pi\mu g R^2(\rho_n - \rho_w)L^2/4$ (5.3.3), where μ is the frictional coefficient between a nanowire and substrate surface with water film on it, g is the gravitational acceleration, R is the nanowire radius, and ρ_n and ρ_w are the density of the nanowire and water, respectively. The frictional coefficient μ can be obtained by tilting the substrate and measuring the critical tilting angle when nanowires in the water film slides with a constant speed. For a 30 μm-long and 100 nm-thick nanowire, μ is measured to be 0.51. Thus the frictional torque T_f on it is estimated to be 8.4×10^{-20} Nm. For the same nanowire, I of the nanowire is calculated to be 1.85×10^{-25} kgm^2. Using the angular acceleration of the transient process shown in Fig. 5.3.5(b), $I\alpha_\theta$ is calculated to be 3.2×10^{-27} Nm, which is much less than T_f. Thus $T_d = T_f$ for both of the transient and steady states, and the driving torque is 8.4×10^{-20} Nm for the AgNW. Based on Eq. 5.3.3, the driving torque has an order of magnitude of 10^{-19} Nm for a 67 μm-long and 100 nm-thick AgNW.

It is known that the acoustic streaming in an ultrasonic field takes on the form of micro eddies in many cases, and the pattern and speed of the micro eddies are well affected by the frequency, boundary and strength of the ultrasonic field. This results in the diversity of acoustic streaming patterns, which provides the possibilities to drive various nanoscale entities with different shape and size. However, before realizing these driving functions, it is necessary to clarify the dependence of acoustic streaming patterns in some relatively complicated ultrasonic fields on the ultrasonic field parameters.

The acoustic streaming field in this method is mobile, that is, it can be generated at any desired location in nanowire suspension by simply shifting the vibrating fiberglass rod to that location. This feature can widen the application range of the method.

In addition to realizing the rotary driving of a single nanowire on a substrate surface, the experimental phenomena and analyses in this work clearly indicate that the acoustic streaming can exist in a plane 100 nm away from a still silicon substrate surface. This means that a still solid surface in acoustic streaming field should be treated as a slip boundary, which is very useful in theoretical calculation of acoustic streaming field.

5.3.4 *Summary*

A strategy to use ultrasonically generated acoustic streaming to rotate a single AgNW in water film on a substrate surface is given. The rotation is always around the nanowire center or end which is directly under the tip of the vibrating fiberglass rod, which is used to excite the acoustic streaming. For a 67μm-long and 100nm-thick AgNW, the revolution speed can be greater than 3 rad/s, and the driving torque has an order of magnitude of 10^{-19} Nm. Measured temperature rise at the manipulation spot is very low (< 0.1℃) in the experiments. The method has potential applications in the orientation and dynamics property measurement of individual nanoscale entities, and assembling of micro/nano structures, etc. The mobility and diversity of the acoustic streaming field make the method competitive in the applications.

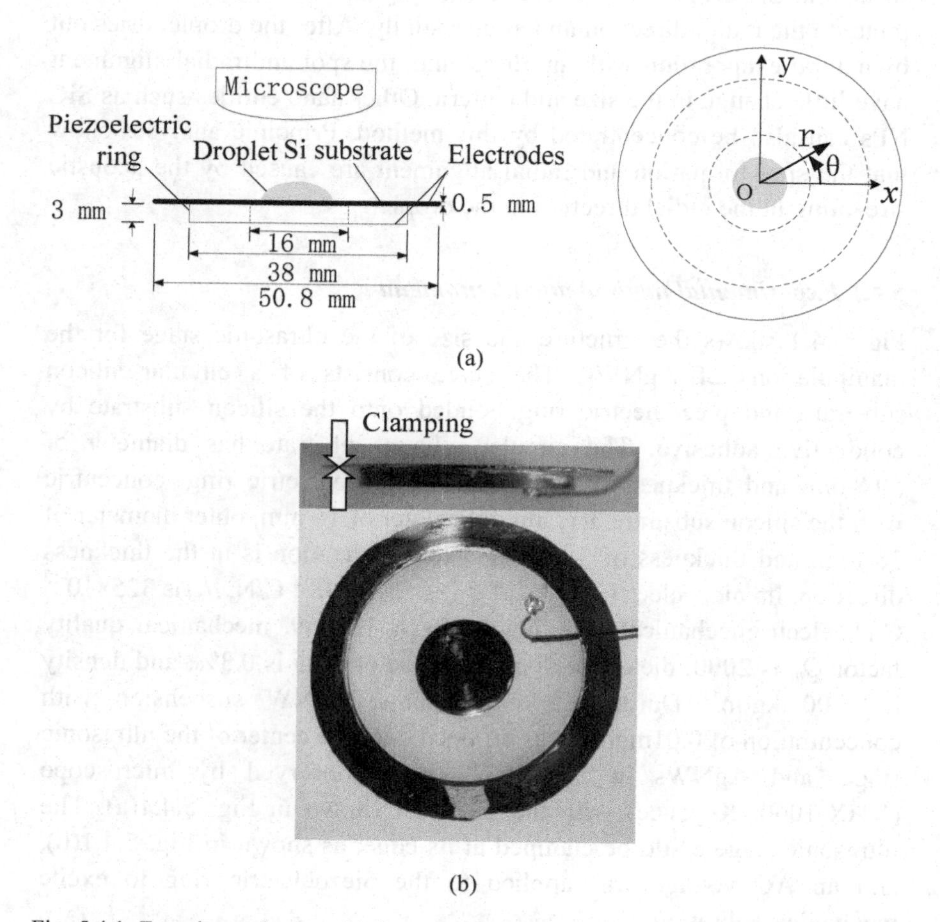

Fig. 5.4.1. Experimental setup. (a) Structure and size of the ultrasonic stage. (b) Images of the ultrasonic stage.

5.4 Concentration and Alignment of Nanoscale Entities

A strategy for the concentration and alignment of nanoscale entities is described in this section [4]. The manipulations are implemented within a nano entity suspension droplet at the center of an ultrasonic stage. When the ultrasonic stage vibrates properly, AgNWs on the substrate surface in the droplet may move to the stage center and form a spot or rotate to the radial direction and align radially. After the droplet dries out by natural evaporation without ultrasound, the spot and radial alignment have little change in the size and pattern. Other nano entities such as SiC NPs can also be concentrated by this method. Principle analyses show that the spot formation and radial alignment are caused by the acoustic streaming in the radial direction in the droplet.

5.4.1 *Experimental method and phenomena*

Fig. 5.4.1 shows the structure and size of the ultrasonic stage for the manipulations of AgNWs. The stage consists of a circular silicon substrate and piezoelectric ring bonded onto the silicon substrate by conductive adhesive. The circular silicon substrate has diameter of 50.8 mm and thickness of 0.5 mm. The piezoelectric ring, concentric with the silicon substrate, has inner diameter of 16 mm, outer diameter of 38 mm, and thickness of 3 mm, and its polarization is in the thickness direction. Its piezoelectric constant d_{31} is -145×10^{-12} C/N, d_{33} is 325×10^{-12} C/N, electromechanical coupling factor k_p is 0.59, mechanical quality factor Q_m is 2000, dielectric dissipation factor $tan\delta$ is 0.3%, and density is 7700 kg/m^3. During the experiments, AgNW suspension with concentration of 0.01mg/ml was dropped onto the center of the ultrasonic stage, and AgNWs in the droplet were observed by microscope (VHX-1000, Keyence) with the direction shown in Fig. 5.4.1(a). The ultrasonic stage could be clamped at its edge, as shown in Fig. 5.4.1(b), and an AC voltage was applied to the piezoelectric ring to excite mechanical vibration.

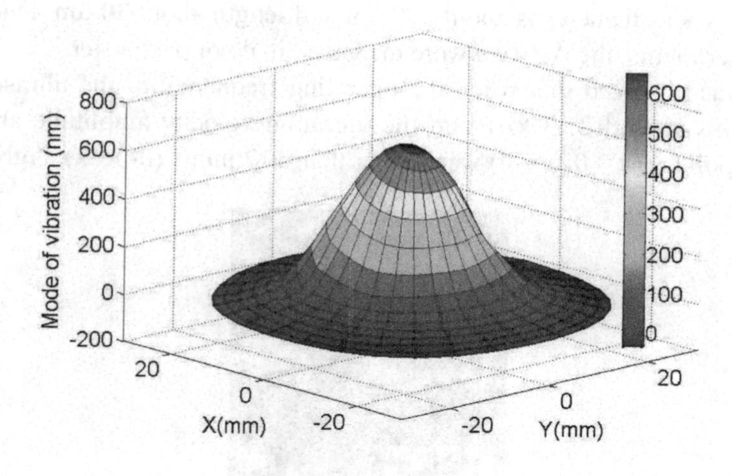

Fig. 5.4.2. Measured vibration displacement distribution of the ultrasonic stage at 21.3 kHz.

The out-of-plane vibration of the upper surface of the ultrasonic stage was measured by laser Doppler vibrometer POLYTEC PSV-300F. It was found that the resonance frequency was 21.3 kHz at 60 Vp-p operating voltage. Measured vibration mode at this frequency is shown in Fig. 5.4.2. It is seen that there is a vibration peak at the stage center o ($x = 0, y = 0$), and the vibration mode of the ultrasonic stage is symmetric about the center approximately. AgNWs used in the experiments are shown in Fig. 5.4.3, which was taken by Scanning Electron Microscope

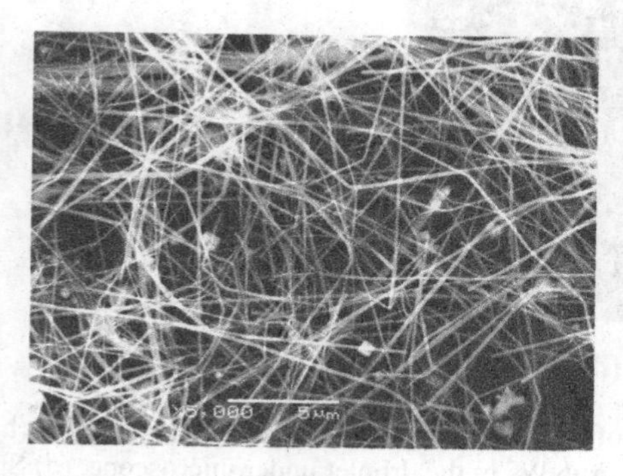

Fig. 5.4.3. AgNWs under SEM (Scanning Electron Microscope).

(SEM). Their diameter is about 100 nm and length about 30 μm. During the experiments, the AgNWs were dispersed in deionized water.

It was observed that when the operating frequency of the ultrasonic stage was around 21.3 kHz and the vibration velocity amplitude at the center point o ($x = 0$, $y = 0$) was higher than 112 mm/s (0-peak), AgNWs

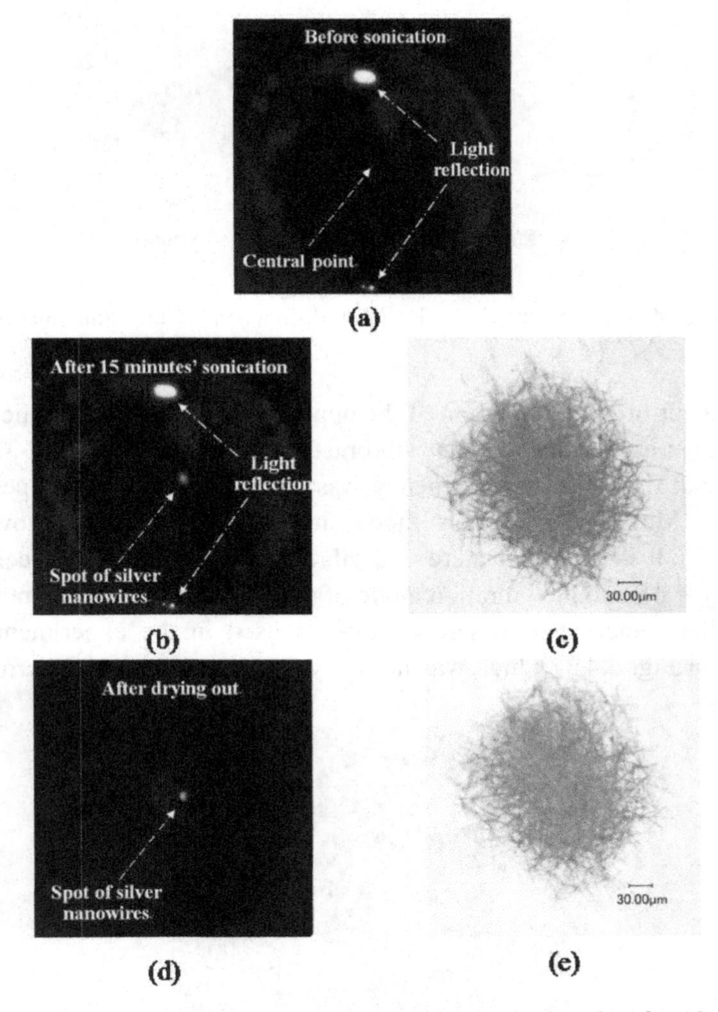

Fig. 5.4.4. Formation of the spot of AgNWs. (a) Before sonication. (b) After 15 minutes' sonication. (c) Spot of AgNWs in the droplet under microscope. (d) Spot of AgNWs after the droplet dries out. (e) Spot of AgNWs under microscope after the droplet dries out.

in the droplet moved to the center of the silicon substrate and a spot was formed at the stage center in less than 15 minutes. Fig. 5.4.4 shows the formation of a AgNW spot in a droplet of 40 μL when the operating frequency is 21.3 kHz and vibration velocity is 144 mm/s (0-p) at central point *o*. Fig. 5.4.4(a) shows that there is no AgNW spot before sonication; Fig. 5.4.4(b) shows that there is a AgNW spot at central point *o* after sonication; Fig. 5.4.4(d) shows that the AgNW spot still exists after the droplet dries out by natural evaporation (without ultrasound). Figs. 5.4.4(a), (b) and (d) were taken by Canon EOS 550D camera. Figs. 5.4.4(c) and (e), taken by microscope (VHX-1000, Keyence), show magnified images of the spots in Figs. 5.4.4(b) and (d), respectively. It is seen that the spot diameter has little change before and after the droplet dries out by natural evaporation. The diameter and thickness of the spot are 204 μm and 40 μm, respectively. The spot is not perfectly circular and the major and minor axes of the spot have length difference of 5%. In this paper, the minor axis of the spot is used as the spot diameter. The thickness of the spot can be measured by the focal distance change when the focal point is shifted from the substrate surface to the spot surface.

It was also observed that when the operating frequency of the ultrasonic stage was 21.3 kHz and the vibration velocity amplitude of central point *o* was higher than 19.6 mm/s (0-p) and lower than 70 mm/s (0-p), the motion of AgNWs to the central point *o* was very slow and it took more than one hour to form the spot. However, in this case, AgNWs in the droplet could rotate to the radial direction (the direction pointing to the stage center), and eventually aligned in the radial direction in less than 15 minutes. Fig. 5.4.5 shows the direction alignment of AgNWs in a droplet of 40 μL when the operating frequency is 21.3 kHz and the vibration velocity is 42 mm/s (0-p) at central point *o*. Fig. 5.4.5(a) shows the AgNWs around point *o* in the droplet after 15 minutes' sonication, and Fig. 5.4.5(b) the AgNWs around point *o* after the droplet in Fig. 5.4.5(a) dries out by natural evaporation. Figs.5.4.5(c) shows the AgNWs around point (r = 1 mm, θ = 210°) in the droplet after 15 minutes' sonication, and Fig. 5.4.5(d) the AgNWs around point (r = 1 mm, θ = 210°) after the droplet in Fig. 5.4.5(c) dries out by natural evaporation. Based on our experimental observation, when the operating frequency of the ultrasonic stage is 21.3 kHz and the vibration velocity at central point o is higher than 70 mm/s (0-p) but lower than 112 mm/s (0-p), the transfer and rotation of AgNWs happen simultaneously.

5.4.2 *Principle analyses*

Due to the vibration velocity peak at central point o, acoustic streaming from the ultrasonic stage center to the top of the droplet occurs. Due to the conservation law in fluid dynamics, the AgNW suspension on the substrate surface flows to the stage center along the radial direction. So the flow in the droplet may be described by Fig. 5.4.6. With the

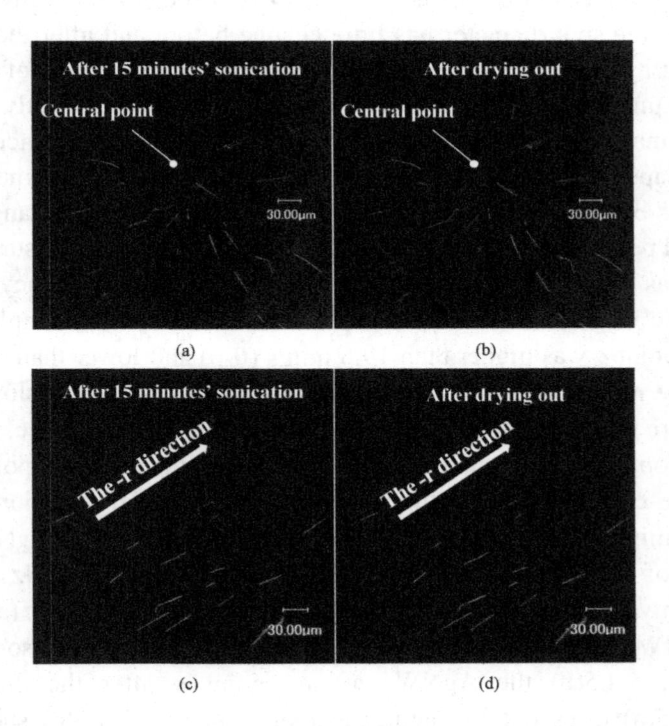

Fig. 5.4.5. The radial alignment of AgNWs. (a) Alignment of AgNWs at the stage center after 15 minutes' sonication. (b) Alignment at the stage center after drying out by natural evaporation. (c) Alignment at r = 1 mm, θ = 210° after 15 minutes' sonication. (d) Alignment at r = 1 mm, θ = 210° after drying out by natural evaporation. The drying process is implemented by natural evaporation after switching off the ultrasound.

suspension of clumps of SiC nanowires, this flow pattern was confirmed by the naked eye. The flow pattern can well explain the AgNW spot formation and radial alignment. AgNWs are brought to the stage center by the radial flow; due to the gravity on the AgNWs, the AgNWs arriving at the stage center cannot move upward with the acoustic streaming; eventually, a spot is formed at the stage center due to the accumulation of AgNWs.

The radial flow also aligns the nanowires in the radial direction. When the ultrasonic stage vibration is large, the transfer speed of AgNWs to the stage center is large, and there is not sufficient time for the AgNWs to complete the alignment before arriving at the stage center, which explains the quite random orientation of AgNWs in Figs. 5.4.4(c) and (e). When the ultrasonic stage vibration is small, the transfer speed to the stage center is slow and there is enough time for the AgNWs to rotate to the radial direction and eventually to align in the radial direction.

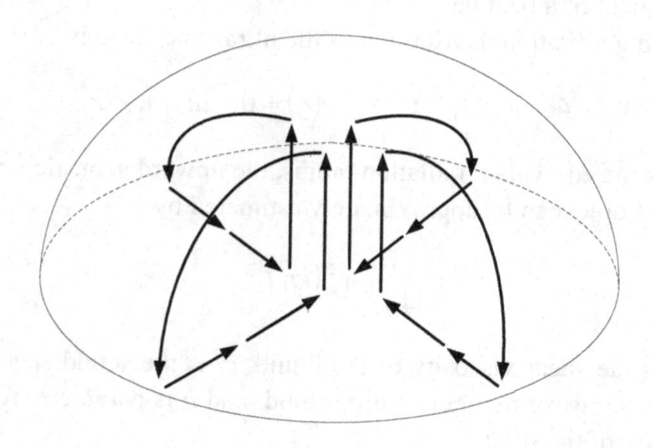

Fig. 5.4.6. Acoustic streaming pattern in the droplet.

The transfer of a bent AgNW on the substrate surface to the stage center was recorded by the microscope, and the images are listed in Fig. 5.4.7(a). In the images, the line on the substrate surface, generated during wafer fabrication, is used as a motion reference, and the arrow represents the transfer direction pointing to the central point *o*. From image *a1* to *a9*, the bent nanowire moves to the reference line first then crosses over it, and finally moves to the stage center. Using a bent nanowire for the images in Fig. 5.4.7(a) is for the convenience of observation. Most nanowires in the droplet are straight, and according to our experimental observation, they can move to the stage center just like the bent one in Fig. 5.4.7(a). The rotation of a AgNW on the substrate surface was recorded by the microscope, and the images to show the rotation are listed in Fig. 5.4.7(b). In the images in Fig. 7.4.7(b), the arrow points to the central point *o*. From image *b1* to *b9*, the nanowire rotates anticlockwise; in image *b9*, the nanowire eventually stops rotating and aligns in the radial direction. According to the experimental observation, a bent nanowire can also rotate but its radial alignment is poor. A bent nanowire has two straight sections. When the driving torque on one of the two sections is zero, the driving torque on another section is not zero due to the angle between the two sections. This causes the poor alignment of a bent nanowire.

Assuming vibration distribution of the ultrasonic stage is

$$v(r) = v_m \quad at \quad r < r_1 \quad ; \qquad v(r) = 0 \quad at \quad r_1 < r \tag{5.4.1}$$

where r_1 is the equivalent radiation radius, the upward acoustic streaming inside the droplet can be approximately estimated by

$$V_z \propto \frac{b}{4\eta c_0} v_m^2 (kr_1)^2 \tag{5.4.2}$$

where η is the shear viscosity of the liquid, c_0 is the sound speed in the liquid, k is the wave number of ultrasound, and b is parameter related to the viscosity of the fluid.

$$b = 4\eta/3 + \eta' \tag{5.4.3}$$

where η is the shear viscosity and η' is the volume viscosity. Considering the continuity of flow, the radial speed of acoustic streaming on the substrate at location r is

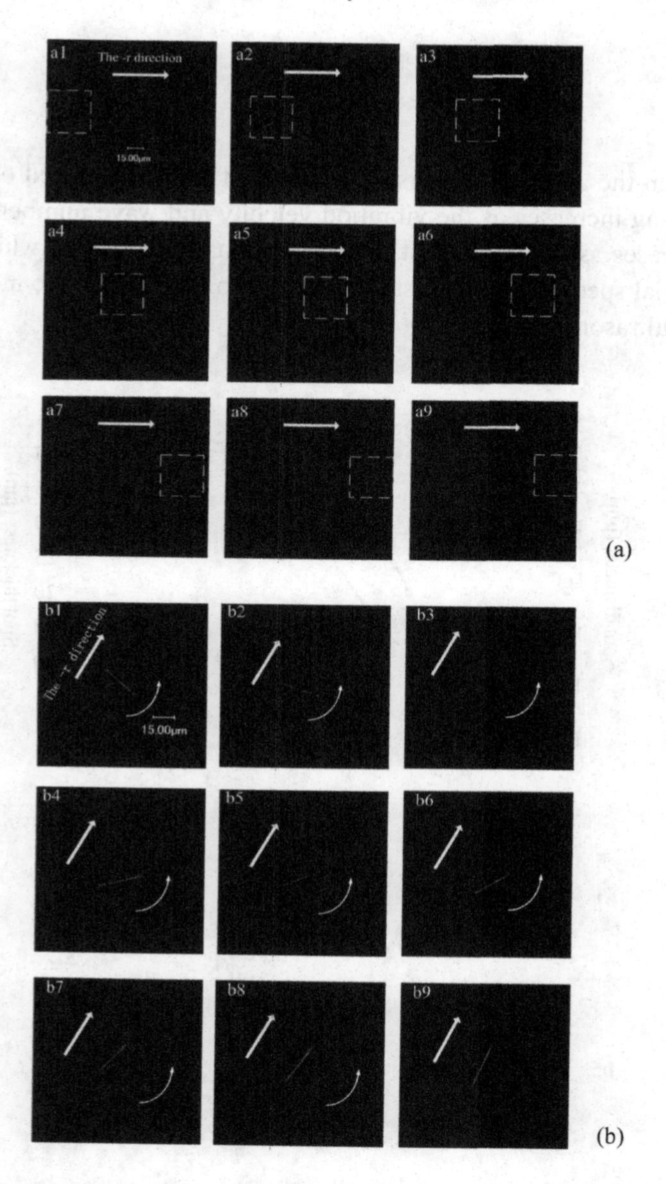

Fig. 5.4.7. Transfer and direction alignment process of the AgNWs in a droplet of 40 μL at 21.3 kHz operating frequency. (a) Transfer of a bent nanowire to the stage center when the vibration velocity is 144 mm/s (0-p) at central point *o*. (b) Rotation of a nanowire to the radial direction when the vibration velocity is 42 mm/s (0-p) at central point *o*. The arrows point to the central point *o*.

$$V_r \propto \frac{V_z r_1^2}{2r} \tag{5.4.4}$$

From the above equations, it is seen that the radial speed of acoustic streaming increases as the vibration velocity and wave number increase; it increases as the equivalent radiation radius r_1 increases, which means the radial speed of acoustic streaming is also affected by the mode shape of the ultrasonic stage.

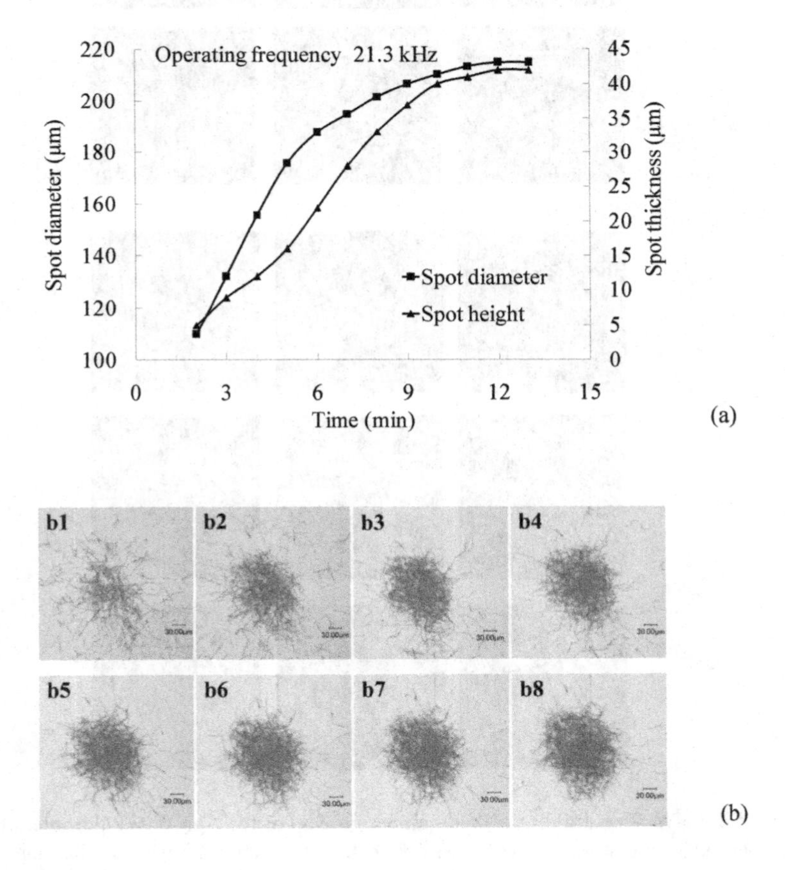

Fig. 5.4.8. Formation of the spot. (a) The spot diameter and thickness vs. sonication time for a droplet of 40 μL. (b) Images to show the spot formation process.

5.4.3 *Characteristics and discussion*

The spot diameter and thickness versus sonication time were measured for a AgNW suspension droplet of 40 μL when the vibration velocity amplitude is 144 mm/s (0-p) at point *o*, and the results are shown in Fig. 5.4.8(a). It shows that the spot diameter and thickness becomes steady when the sonication time is long enough. This indicates that when the sonication time is long enough, there are little AgNWs moving to the center and all of the AgNWs which the radial acoustic streaming can affect have moved to the stage center. The images in Fig. 5.4.8(b) show the spot formation process, in which the time interval between two adjacent images is one minute.

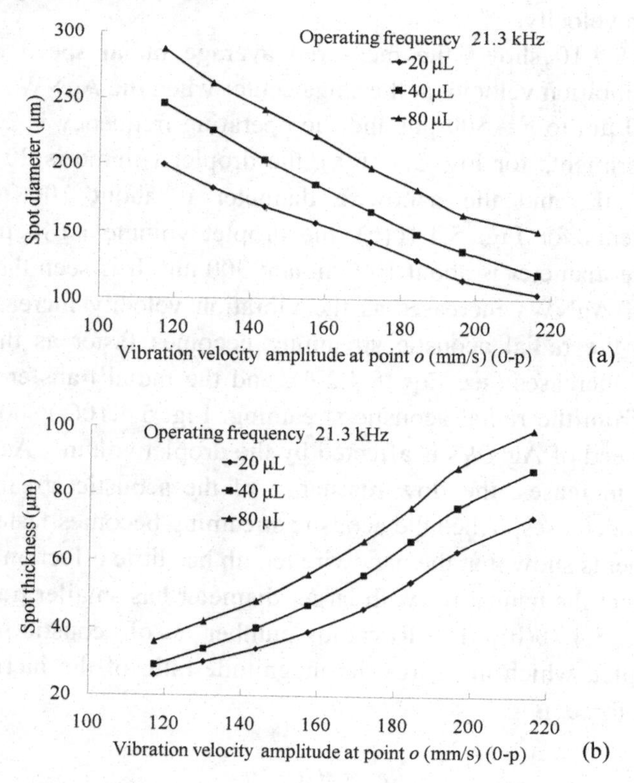

Fig. 5.4.9. The spot diameter and thickness vs. vibration velocity at point *o* (0-peak) for the droplets of 20 μL, 40 μL and 80 μL. (a) Spot diameter. (b) Spot thickness.

The spot diameter and thickness of steady state vs. vibration velocity amplitude at central point *o* was measured for AgNW suspension droplets of 20 μL, 40 μL and 80 μL at 21.3 kHz operating frequency, and the results are shown in Fig. 5.4.9. From Fig. 5.4.9, it is seen that the spot diameter decreases and the spot thickness increases as the vibration velocity increases, and the spot diameter and thickness also depend on the droplet volume. As the vibration at central point *o* increases within the experimental range, the vibration velocity inside the droplet increases. This can increase the speed of the radial acoustic streaming inside the droplet, and thus results in the characteristics shown in Fig. 5.4.9. As the droplet volume increases, there are more AgNWs in a droplet, which causes larger spot diameter and thickness for a given vibration velocity.

Fig. 5.4.10 shows the measured average radial speed of AgNWs versus vibration velocity at the stage center when the AgNWs move from r = 1000 μm to r = 500 μm and the operating frequency is 21.3 kHz. In the experiments for Fig. 5.4.10(a), the droplet volume is 20 μL, 40 μL and 80 μL, and the nanowire diameter is about 100 nm. In the experiments for Fig. 5.4.10(b), the droplet volume is 40 μL, and the nanowire diameter is about 100 nm and 300 nm. It is seen that the radial speed of AgNWs increases as the vibration velocity increases. This is because the radial acoustic streaming becomes faster as the vibration velocity increases (see Eqs. 5.4.2-4), and the radial transfer of AgNWs results from the radial acoustic streaming. Fig. 5.4.10(a) shows that the radial speed of AgNWs is affected by the droplet volume. As the droplet volume increases, the flow resistance of the acoustic streaming in the droplet decreases, hence the acoustic streaming becomes faster. Also, the experiments show that the nanowire length has little effect on the transfer speed, and the nanowires with larger diameter has smaller transfer speed [see Fig. 5.4.10(b)]. The Reynolds number *Re* of acoustic streaming in the droplet, which measures the magnitude ratio of the inertial force to viscous force, is

$$Re = DU\rho/\eta \tag{5.4.5}$$

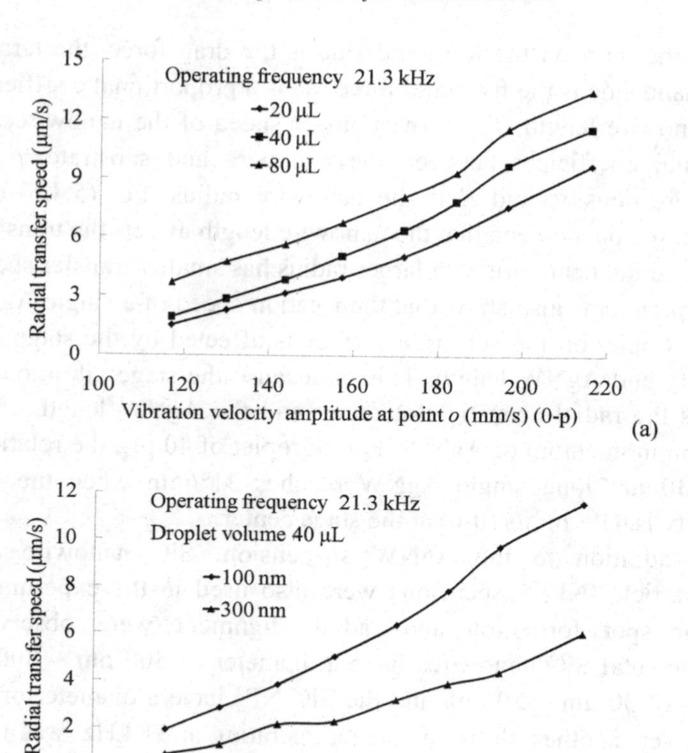

Fig. 5.4.10. Radial transfer speed of AgNWs vs. vibration velocity at point *o* (0-peak). (a) For three different droplet volumes. (b) For two different nanowire diamters.

where D is the nanowire diameter, U is the relative velocity between the nanowire and streaming, and ρ is the density of water. In the experiments, it is less than 10^{-5}. Thus for a nanowire in the droplet, the inertia force can be neglected in force analyses, and the drag and frictional forces on a nanowire are in balance. There is

$$K_1 \eta L(V_r - V_{nw}) = \mu_f (\rho_{nw} - \rho)\, L(\pi R^2)\, g \qquad (5.4.6)$$

where the term on the left hand side is the drag force, the term on the right hand side is the frictional force, K_l is a proportional coefficient, L is the nanowire length, V_{nw} is the transfer speed of the nanowire, μ_f is the frictional coefficient between the nanowire and substrate, ρ_{nw} is the nanowire density, and R is the nanowire radius. Eq. (5.4.6) can well explain the phenomena that the nanowire length affects the transfer speed little, and the nanowire with larger radius has smaller transfer speed.

Experiments also show that the rotation speed of a single AgNW in a micro droplet on the substrate surface is affected by the stage vibration velocity and AgNW length. This is because the stage vibration velocity affects the radial acoustic streaming, and the AgNW length affects the rotation momentum of AgNW. For a droplet of 40 μL, the rotation speed of a 30 μm long single AgNW reaches 31°/min when the vibration velocity is 41.4 mm/s (0-p) at the stage center.

In addition to the AgNW suspension, SiC nanowire and SiC nanoparticle (NP) suspensions were also used in the experiments, and similar spot formation and radial alignment were observed. The experimental SiC nanowires have a diameter of 500 nm ~ 800 nm and length of 30 μm ~ 50 μm, and the SiC NPs have a diameter of 400 nm. Moreover, another ultrasonic stage, operating at 31 kHz, was fabricated by a 0.2 mm thick copper plate and piezoelectric ring (PZT8) of 12 mm (outer diameter) × 6 mm (inner diameter) × 1.2 mm (thickness), with the same structural topology as shown in Fig. 5.4.1. AgNWs, SiC NWs and SiC NPs in the experimental suspension droplets could transfer to the stage center to form a spot; the AgNWs could align in the radial direction at proper stage vibration; the SiC NWs and their agglomerations could also align in the radial direction at proper stage vibration. Fig. 5.4.11 is an image to show a spot formed by the SiC NWs at the vibration peak position of the copper plate. The measured spot diameter is 1 mm, and vibration velocity at the spot is 37.8 mm/s (0-p). Thus the spot formation and radial alignment method proposed by this work may be applied to other nano materials.

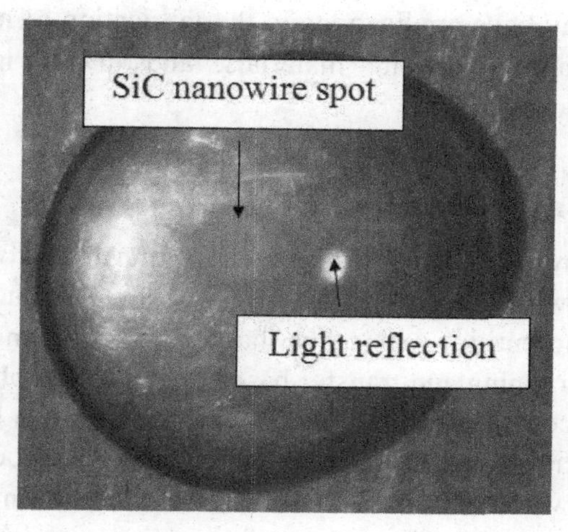

Fig. 5.4.11. Ultrasound induced spot of SiC nanowires in a droplet on a 31 kHz ultrasonic stage with copper substrate.

5.4.4 *Conclusions*

It is demonstrated that nanowires in a micro liter droplet located at the center of a low-frequency circular ultrasonic stage can transfer to the stage center to form a spot or rotate to the radial direction, if the stage vibration is proper. The ultrasonic stage vibrates symmetrically about its center, with a vibration peak at the center. As the ultrasonic stage vibration increases, the spot diameter deceases and spot thickness increases. Under the experimental conditions, the spot diameter and thickness are several hundred and several ten microns, respectively. Switching off the ultrasound and drying out the droplet by natural evaporation, the spot and alignment have little change in the size and pattern. The acoustic streaming in the radial direction in the droplet is responsible for the AgNW spot formation and radial alignment. Other nano entities can also be concentrated by this method.

The method proposed in this paper has the following merits: the device is simple and easy to use; the technique has the potential to deal with large volume droplets of nano entity suspensions and to construct an array of ultrasonic stages for mass fabrication. The technology proposed by this work may have applications in the fabrication of nano electronic devices and nano composite materials, and in the high-sensitivity detection of nano substance.

5.5 Summary and Remarks

The acoustic streaming can be used to trap, transfer, drive, concentrate and align nanowires in water film on a substrate surface. It can also concentrate nano entities with other shapes in water film on a substrate surface. In the trapping and transfer based on a micro probe in ultrasonic vibration, the acoustic streaming is mobile, that is, it can be moved with moving the micro probe. Thus these manipulations can be carried out at any location in water film on a substrate. The vibration mode and shape of the micro probe affect the pattern of acoustic streaming around itself. Contact and noncontact type trapping can be realized by using microprobes with different vibration modes and shapes. In the concentration of nanoscale entities based on an ultrasonic stage, the size and shape of the spot formed by the concentration process have little change after the water film dries.

Based on the manipulation principles shown by the examples in this chapter, more nano manipulation functions can be developed for particular applications in the fabrication of microelectronic/photonic devices, micro/nano assembling, high-sensitivity bio-sensing, etc.

References

1. Li, N., Hu, J. H., Li, H. Q., Bhuyan, S. and Zhou, Y. J. (2012). Mobile Acoustic Streaming Based Trapping and 3–Dimensional Transfer of a Single Nanowire, Appl.Phys. Lett., 101 (9), 093113.
2. Li, H. Q. and Hu, J. H. (2103). Noncontact Manipulations of a Single Nanowire Using an Ultrasonic Micro-Beak, IEEE T Nanotechnol., in press.
3. Li, N., Hu, J. H. (2103). Sound Controlled Rotary Driving of a Single Nanowire, IEEE T Nanotechnol., in press.
4. Zhou, Y. J., Hu, J. H. and Bhuyan, S. (2013). Manipulations of Silver Nanowires in a Droplet on Low–Frequency Ultrasonic Stage, IEEE Trans. Ultrason. Ferroelectr. Freq. Control, 60 (3), pp. 622–629.

Chapter 6

Ultrasonic Microfluidic Manipulations

This chapter gives the examples of ultrasound based microfluidic manipulations. The contents include a lobed pattern in a film of suspension, adsorption and extraction of droplets without a chamber, microfluidic transportation in and droplet generation on a twisted bundle of metal wires, droplet generation and rotary driving with an ultrasonic hollow needle, and merging of droplets. The examples in this chapter help one better understand the physical principles employed in ultrasonic microfluidic manipulations.

6.1 Lobed Pattern in a Film of Suspension

It is known that small particles in acoustic field may form some particular patterns. They include Chladni's patterns and Kundt's patterns. Chladni's patterns are generated by small solid particles moving to the nodal lines, circles and points of a vibrating plate. Kundt's patterns are generated by small solid particles moving to the nodal planes of longitudinally vibrating air column in a tube. Stationary sound waves are used in the generation of these patterns. The physical effect of forming these patterns has been used to visualize the vibration mode pattern, and to separate and filter particles from suspension.

A lobed pattern around an ultrasonically vibrating needle in aqueous suspension film of micro particles is demonstrated in this section [1]. The contents of this section include experimental setup and phenomenon, physical principle, characteristics, and discussion.

6.1.1 *Experimental setup and phenomena*

Figure 6.1.1 shows the experimental setup to generate and observe a lobed pattern in aqueous suspension film of yeast particles. A layer of aqueous suspension film of yeast particles with a thickness of about 1.0 mm is dispersed on a glass slide, and a vibrating stainless steel needle is inserted into the suspension film horizontally, as shown in Figs. 6.1.1(a) and (b). The suspension film is observed by microscope (BX51, Olympus), and the lobed pattern is recorded by common high pixel digital camera. The stainless steel needle is welded onto the side face of one of the rectangular plates in a sandwich piezoelectric transducer, as shown in Fig. 6.1.1(c). In the piezoelectric transducer, four piezoelectric rings are aligned and pressed by a bolt-plate structure, with the poling directions and electrode configuration shown in Fig. 6.1.1(c). The outer diameter, inner diameter and thickness of each piezoelectric ring are 20 mm, 12 mm and 2.4 mm, respectively. The electromechanical quality factor Q_m, piezoelectric coefficient d_{33}, and relative dielectric constant $\varepsilon_{33}^T / \varepsilon_0$ of the piezoelectric rings are 2000, 325×10^{-12} m/v and 1450, respectively. The size of each stainless plate at the two ends of the transducer is $20\times20\times2.3$ mm^3. The needle has length of 27.8 mm from the edge of the stainless steel plate, and diameter of 0.83 mm which gradually decreases down to its tip from z = 11.5 mm. The length of the needle welded onto the transducer is 7 mm. Measured resonance frequency of the transducer is 64.4 kHz for input voltage of 35.55 Vrms. At the resonance, the piezoelectric transducer shown in Fig. 6.1.1(c) utilizes the thickness vibration of the piezoelectric stack to excite a flexural vibration mode of the needle in the rz plane; thus in Fig. 6.1.1(a), the needle vibration is perpendicular to the page and parallel to the suspension film. The needle is in contact with the substrate except its tapered section.

The house hold baking yeast particles (Saccharomyces Cerevisiae) are used in the experimental aqueous suspension. Particle concentration of the suspension is 0.87 mg/ml and the diameter of yeast particles is

4~6 microns. Fig. 6.1.2 shows an image of the yeast particles dispersed in the suspension before sonication. The average number of yeast particles in a square 200 μm×200 μm is about 175.

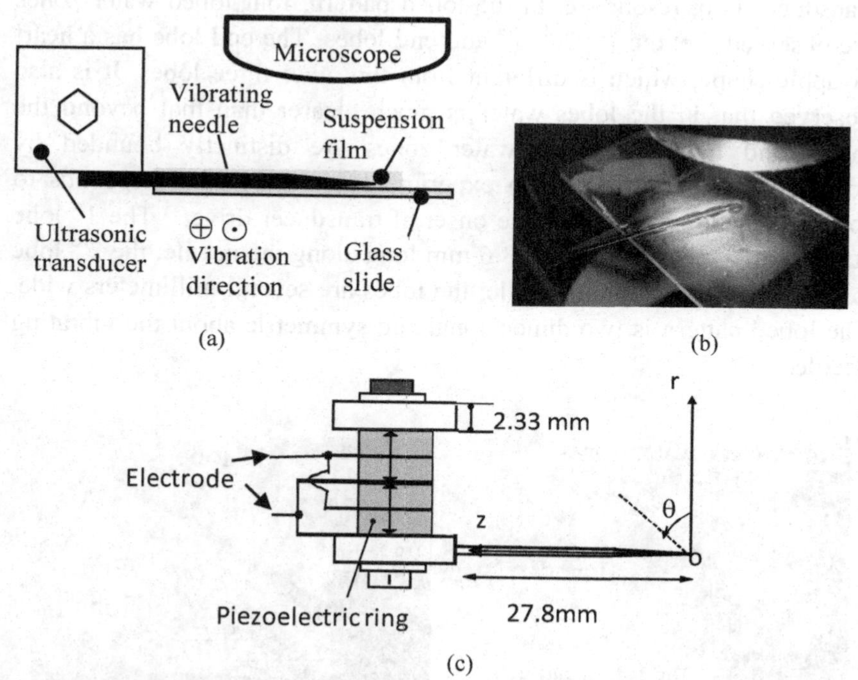

Fig. 6.1.1. Experimental setup for generation and observation of the lobed patterns in an aqueous suspension film of yeast particles on a glass slide. (a) Schematic diagram. (b) Photo (c) Construction of the ultrasonic transducer [top view of the transducer in (a)].

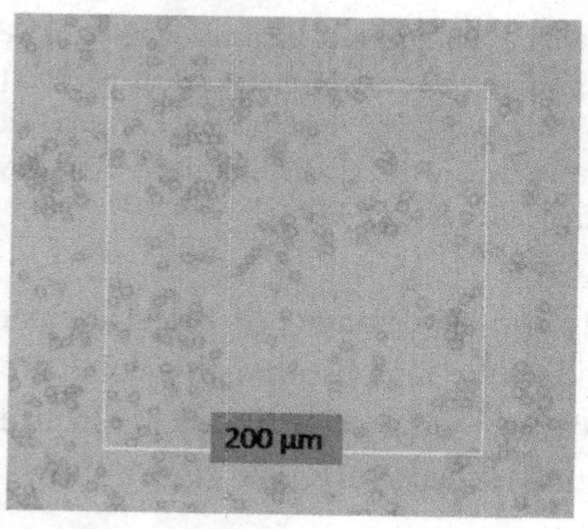

Fig. 6.1.2. Aqueous suspension of yeast particles for experiment.

Fig. 6.1.3 shows the observed lobed patterns in the aqueous suspension film of yeast particles, generated by ultrasonic vibration of the stainless steel needle, at different driving voltages when the transducer is in resonance. In the lobed pattern, four lobed water zones are observed, i.e. the 1st, 2nd, 3rd and end lobes. The end lobe has a heart or apple shape, which is different from the other three lobes. It is also observed that in the lobes water is much clearer than that beyond the lobes, and the four lobed water zones are distinctly bounded by continuous white arc lines. In experiments, it takes 20~30 seconds to form the lobed pattern after the onset of transducer drive. The 1st lobe starts from z = 2.5 mm and is 3.6 mm long along the needle; the 2nd lobe is 4.7 mm long along the needle; the lobes are several millimeters wide. The lobed pattern is two dimensional and symmetric about the vibrating needle.

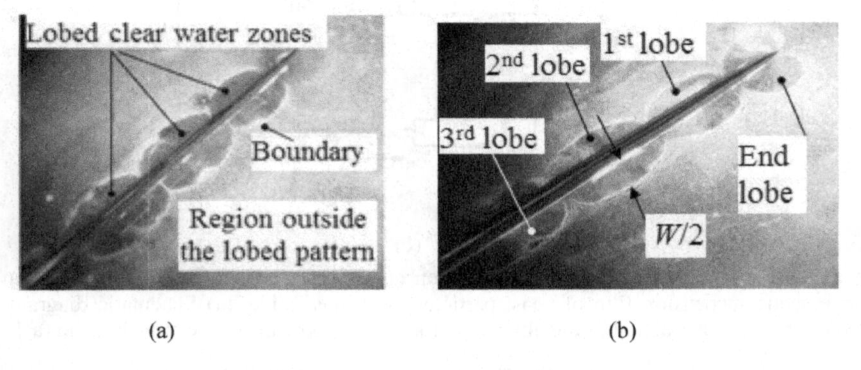

Fig. 6.1.3. Lobed patterns in an aqueous suspension film of yeast particles on a glass slide, generated by ultrasonic vibration of the stainless needle. (a) The input voltage of ultrasonic transducer is 51.4 Vrms, and the transducer operates at resonance frequency of 62.7 kHz. (b) The input voltage of ultrasonic transducer is 53.5 Vrms, and the transducer operates at resonance frequency of 62.7 kHz.

Fig. 6.1.4 gives zoom-in images of the suspension in the 1st lobe, at the boundary of this lobe and beyond this lobe. In the experiment, the operating frequency and voltage of the piezoelectric transducer are 64.4 kHz and 35.55 Vrms, respectively. The figure shows that the water in the

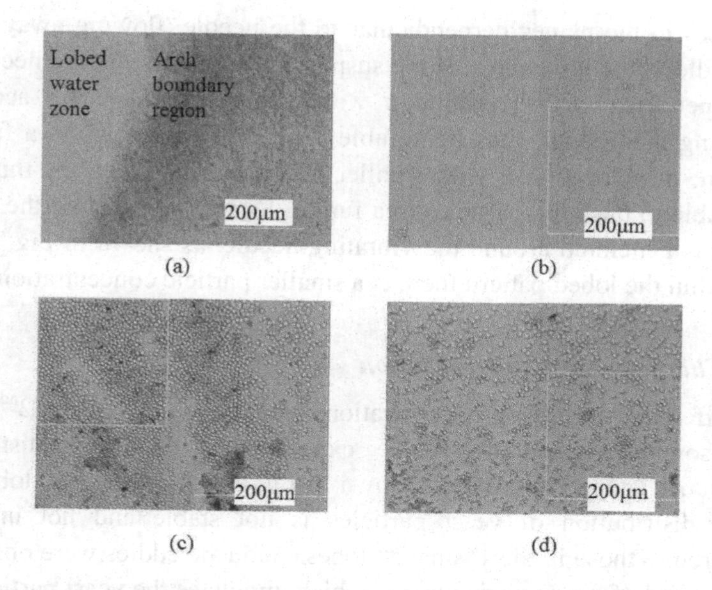

Fig. 6.1.4. Images of the suspension irradiated by ultrasound. The operating frequency and voltage of ultrasonic transducer is 62.8 kHz and 35.55 Vrms, respectively, and the transducer operates in resonance. (a) Image of the clear water zone and boundary. (b) Zoom-in image of the suspension in the 1st lobe. (c) Zoom-in image of the suspension at the boundary of the 1st lobe. (d) Zoom-in image of the suspension beyond the 1st lobe. The measurement point is 0.5 mm away from the boundary.

lobed zone is much clearer than that at and beyond the boundary. From Figs. 6.1.4(b), (c) and (d), it is found that average particle number in a square of 200 μm×200 μm is 21 within the 1st lobe, 625 at the boundary of the 1st lobe, and 250 beyond the 1st lobe (0.5 mm away from the boundary region). The particle concentration at boundary is 30 times of that within the lobe, and 3.6 times of that before sonication.

6.1.2 *Principle analyses*

Excited by the Langevin transducer, the needle vibrates flexurally in the direction parallel to the substrate [2]. An acoustic streaming field can be observed around the needle. It exists on the two sides of the needle, and is approximately symmetric about the needle. The acoustic streaming circulates in the planes perpendicular to the needle, flowing away from the needle in the lower area of the suspension film and into the needle in the upper area. At a position z with larger vibration, the acoustic streaming is stronger, thus being able to push the particles to a further location; at a location z with smaller vibration, it is weaker, thus not being able to push the particles to a further location. Therefore the lobed pattern is generated around the vibrating needle, as shown in Fig. 6.1.3, and within the lobed pattern there is a smaller particle concentration.

6.1.3 *Characteristics and discussion*

The difference of particle concentrations among the end, 1st and 2nd lobes are also investigated under the experimental conditions listed in Section 6.1.1, and the result is shown in Fig. 6.1.5. In the end lobe, the spatial distribution of yeast particles is not stable and not uniform compared to those in the 1st and 2nd lobes. Acoustic eddies were observed near the end of the acoustic needle, which circulates the yeast particles in the end lobe and nearby. The needle has a larger vibration at its sharp end than at other parts, which makes acoustic eddies easy to occur near the sharp end. From Figs. 6.1.5(b) and (c), it is found that particle concentration in the 2nd lobe is 1.7 times of that in the 1st lobe. This may be caused by the larger vibration of the acoustic needle within the 1st lobe, which is closer to the end of the needle.

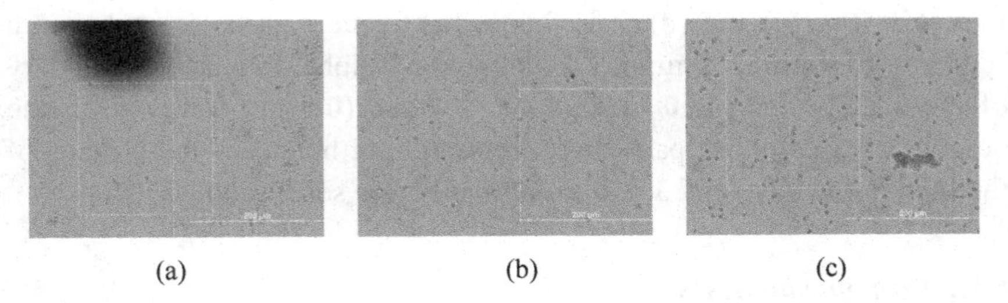

Fig. 6.1.5. Zoom-in images of the suspension in (a) the end lobe, (b) the 1st lobe, and (c) the 2nd lobe. The size of the squares is 200μm×200μm.

The ratio of particle concentration in the 1st lobe to that before sonication (C_c/C_n), was measured for different driving voltages when the transducer was in resonance and the result is shown in Fig.6.1.6. Here C_c/C_n is assumed to be equal to the ratio of the yeast particle number in a 200 μm × 200 μm square in the 1st lobe under sonication to that in the suspension before sonication. The yeast particle number is counted after the lobed pattern becomes stable. From the figure, it is seen that as the vibration increases, the water in the 1st lobe becomes clear first and then turbid. In the figure, the minimum value of C_c/C_n is 6%. The water in the 1st lobe becomes clear and then turbid because the circulation of the acoustic streaming becomes stronger as the vibration increases. It should be pointed out that when the vibration is too large, the lobed pattern becomes very faint. Fig. 6.1.7 shows image of the lobed pattern when the driving voltage of the transducer in resonance is 63.87 V_{rms} and its operating frequency is 61.9 kHz. The end and 1st lobes can be hardly seen in the image. In this case, strong acoustic eddies are seen in the end and first lobes, which circulate the yeast particles in the end and first lobes.

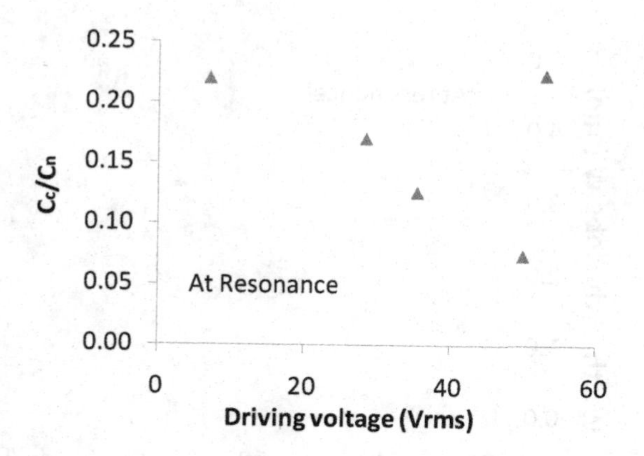

Fig. 6.1.6. Ratio of particle concentration in the 1st lobe under sonication to that before sonication for different driving voltages of the transducer in resonance.

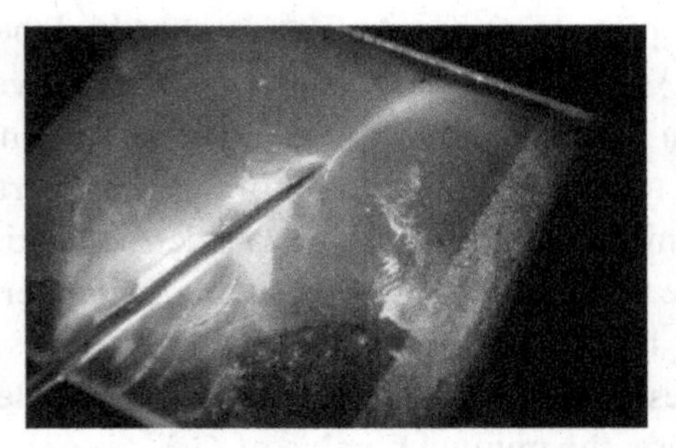

Fig. 6.1.7. Ambiguous lobed pattern around the ultrasoniacally resonating needle when the driving voltage of the transducer in resonance is 63.9 Vrms and its operating frequency is 61.9 kHz.

The width W of the lobed water zones when the ultrasonic transducer was in resonance, was measured for different driving voltages and the results are shown in Fig. 6.1.8. Here, the width W does not include the diameter of the needle. It is seen that the width of the lobed water zones increases with the increase of vibration, which results from increase of the acoustic streaming. In the experiment, as the driving voltage increases from 35.55 Vrms to 63.9 Vrms, the resonance frequency decreases from 64.4 kHz to 61.9 kHz.

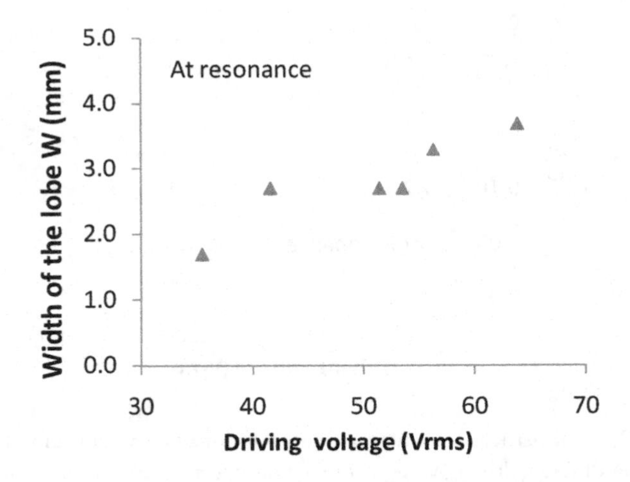

Fig. 6.1.8. The width W of the 1st lobed water zone at different driving voltages when the ultrasonic transducer is in resonance.

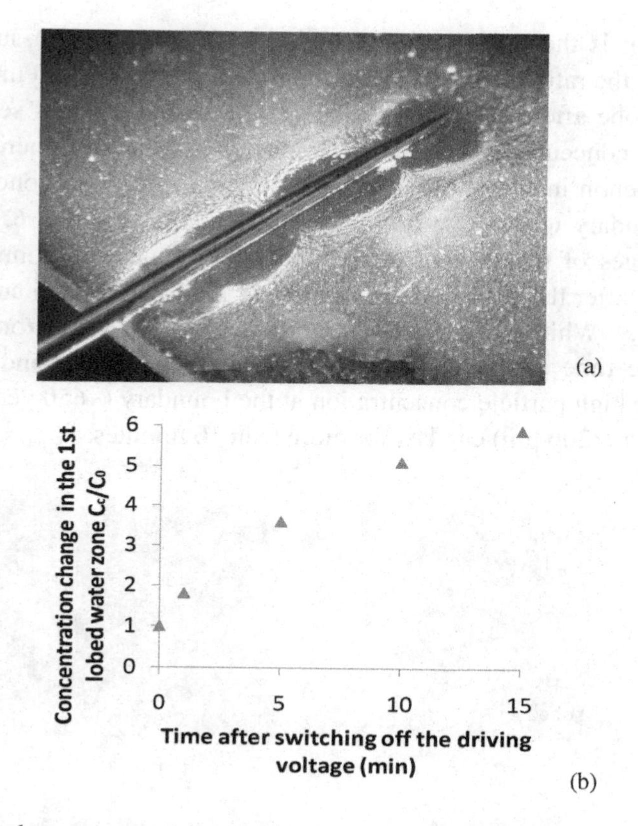

Fig. 6.1.9. Lobed pattern around the needle and particle concentration change in the 1st lobed water zone after sonication is ceased. (a) Image of the lobed pattern 5 minutes after switching off the driving voltage. (b) The particle concentration versus the time after sonication is ceased. During the sonication, the driving voltage and frequency of transducer are 35.55 Vrms and 62.4 kHz, respectively.

Change of the lobed pattern after the needle vibration was ceased, was investigated. Fig. 6.1.9(a) shows an image of the lobed pattern 5 minutes after ceasing the ultrasonic vibration that is excited by driving voltage of 35.55 Vrms and frequency of 62.4 kHz. A lobed pattern can still be seen 5 minutes after the sonication. The normalized particle concentration C_c/C_n in the first lobe after ceasing the vibration versus time period between the measurement instant and sonication end was measured, and the result is shown in Fig. 6.1.9(b). The sonication condition is the same as that for Fig. 6.1.9(a). C_c/C_n is assumed to be equal to the ratio of yeast particle number in a 100 μm×100 μm square in the 1st lobe after sonication to that before sonication. It is seen that the particle concentration increases gradually as the time increases. This phenomenon indicates that it takes time for the particles concentrated in the boundary to disperse into the lobed water zones. Fig. 6.1.10 shows the images of the suspension 1 minute, 5 minutes, 10 minutes and 15 minutes after the ultrasound is ceased, for the same driving conditions as Fig. 6.1.9. While particle concentration in the lobed water zone increases with the time period between the measurement instant and sonication end, the high particle concentration at the boundary (~650 yeast particles in 200 μm×200 μm) can last for more than 10 minutes.

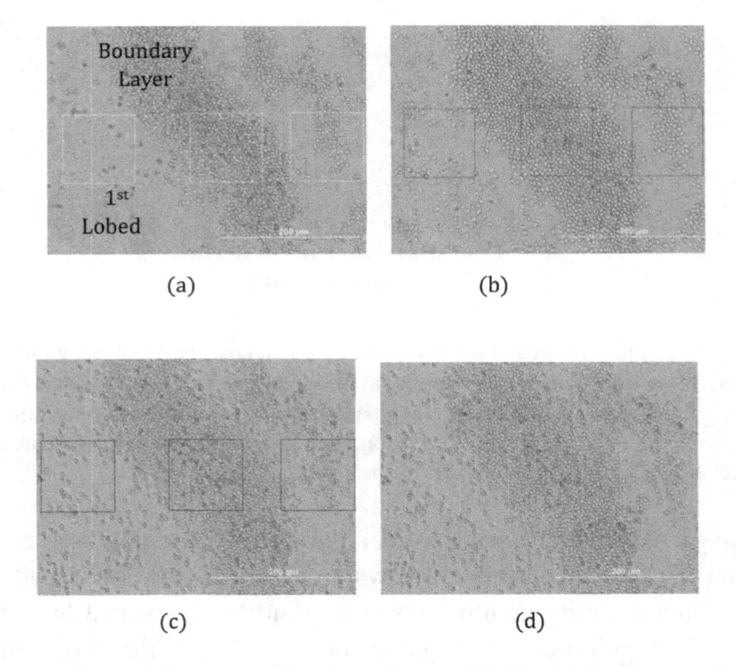

Fig. 6.1.10. Change of yeast particle suspension near the boundary after switching off the ultrasound. (a) 1 minute. (b) 5 minutes. (c) 10 minutes. (d) 15 minutes. The square in the figures has a dimension of 100 μm×100 μm.

6.1.4 *Summary*

A lobed pattern in an aqueous suspension film of micro particles on a glass slide, generated by an ultrasonically vibrating needle, has been demonstrated. Being different from the Chladni's and Kundt's patterns, the acoustic pattern reported in this paper is generated by the acoustic streaming. While the particle concentration in the lobed water zones can be reduced to 6% of that of the suspension before sonication, its value at the lobed pattern boundary is 3.6 times of the suspension before sonication. The lobed pattern can exist for more than ten minutes after ceasing the sonication. Potential applications of this effect include bio sensing, vibration visualization, small-scale suspension purification, aesthetic science, etc.

The experiments which were done by the author's research group, also show that the pattern of acoustic streaming field around the needle depends on the distance between the needle and substrate. When the distance is larger, the acoustic streaming on the substrate surface flows into the location under the needle from the outside, which can concentrate the particles in the lobed areas, rather than clearing the particles from the lobed areas.

6.2 Adsorption and Extraction of Droplets without a Chamber

Ultrasound controlled adsorption and extraction of droplets without a chamber is demonstrated in this section [2]. The contents of this section include the experimental setup, principle analyses, characteristics and discussion.

6.2.1 *Experimental setup*

The structure and size of ultrasonic transducer used to mechanically drive a stainless needle are shown in Fig. 6.2.1(a), as well as the needle size. The transducer has a bolted sandwich structure with four pieces of piezoelectric rings aligned and stacked in the middle, and two stainless plates at the two ends. The four piezoelectric rings have the same dimensions and properties, poled in the thickness direction. Their size is

12 mm (outer diameter) × 6 mm (inner diameter) × 1.2 mm (thickness), piezoelectric constant d_{33} is 250×10^{-12} C/N, electromechanical coupling factor k_{33} is 0.63, mechanical quality factor Q_m is 500, dielectric dissipation factor $tan\delta$ is 0.6%, and density is 7450 kg/m^3. The tightening torque of the transducer is 10 Nm. The stainless needle is welded onto the side surface of one of the two square plates. The needle is 34.8 mm long totally, with 27.8 mm outside the vibration excitation section and 11.5 mm in the tapered section. Two stainless needles are used in the experiments, and their diameters are 0.72 mm and 0.83 mm (above

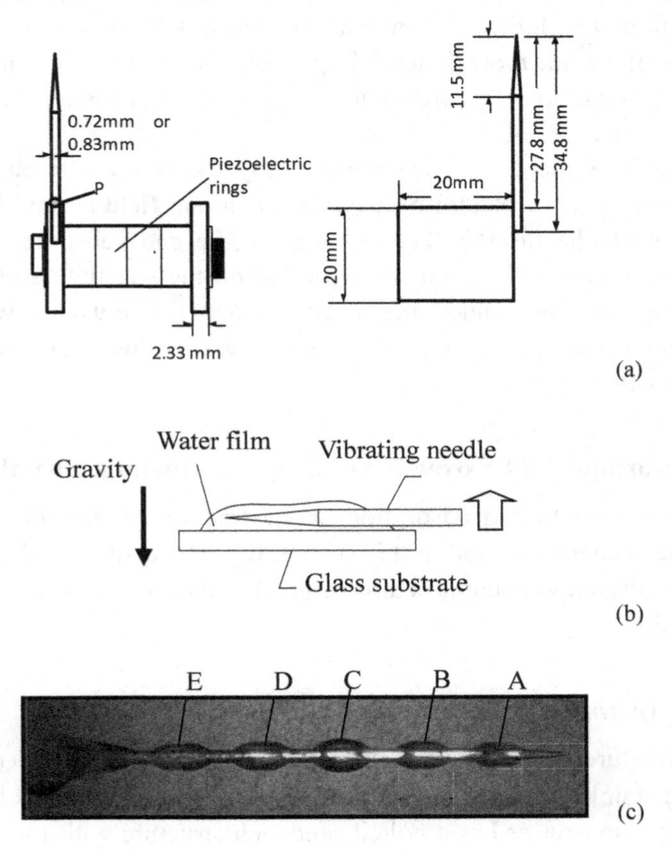

Fig. 6.2.1. Experimental setup and phenomenon. (a) The structure and size of ultrasonic transducer with a solid stainless needle; (b) schematic diagram of experimental setup; (c) image of water droplets adsorbed onto the needle with a diameter of 0.83 mm in vibration.

the tapered section), respectively, and about 0.2 mm at the needle tip. The two stainless plates have the identical size of 20 mm × 20 mm × 2.33 mm. Resonance frequency of the ultrasonic transducer is 77.31 kHz at driving voltage of 10 $V_{p\text{-}p}$. The experimental setup is shown in Fig. 6.2.1(b). A thin water film about 1 mm thick is dispersed on a glass substrate. The needle is horizontally submerged into the water film, with its vibration direction parallel to the substrate. Tuning the needle vibration properly and lifting it from the water film into air, it can be observed that some water droplets adsorb on the needle surface in vibration, with the pattern shown in Fig. 6.2.1(c). In Fig. 6.2.1(c), the needle diamter is 0.83 mm, and the center of droplets *A, B, C, D* and *E* is 3.8 mm, 8.1 mm, 12.6 mm, 17.1 mm and 21.6 mm away from the needle tip respectively. It is experimentally confirmed that when the needle has no vibration, no droplet adsorbs on it after being lifted. Therefore ultrasonic vibration of the needle causes the adsorption of water droplets.

6.2.2 *Operating principle*

Sound pressure *p* can cause a density change of liquid acoustic medium, which is $\Delta\rho = p/c^2$, where *c* is the speed of sound in the medium. During the negative half circle of sound pressure, negative sound pressure results in the density decrease of the medium, and increases the distance among the molecules. In this case, cohesive force among the molecules becomes small. During the positive half circle of sound pressure, positive sound pressure compresses the molecules in the medium. But this cannot increase the cohesive force too much because the molecules repel each

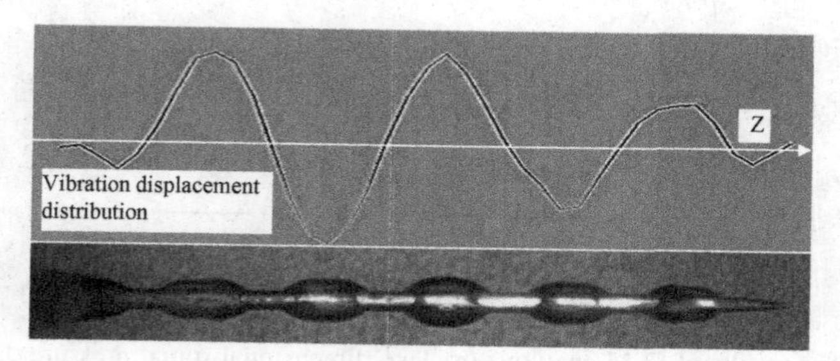

Fig. 6.2.2. Comparison of the measured distributions of needle vibration and adsorbed water droplets.

other when they are too close. Therefore sound pressure makes the cohesive force among the molecules of liquid acoustic medium weak on the time average.

The distribution of needle vibration parallel to the substrate along the length direction was measured by laser Doppler vibrometer (POLYTEC PSV-300F), and the result is shown in Fig. 6.2.2. Images of the adsorbed water droplets are inserted into Fig. 6.2.2 for comparison. It is seen that the centers of adsorbed droplets are at the anti-nodal points of the needle vibration. After that, two dimensional sound field around the vibrating needle is calculated by the FEM (COMSOL MULTIPHYSICS). In the calculation, the needle size shown in Fig. 6.2.1(a) is used, and the needle

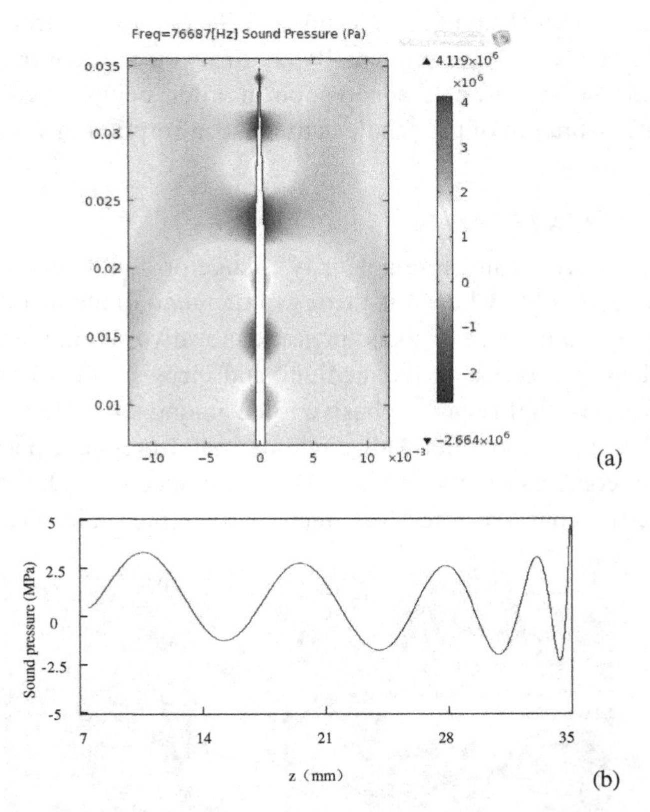

Fig. 6.2.3. The results of FEM analyses. (a) Two dimensional sound pressure distribution around the vibrating needle; (b) The distribution of sound pressure on the surface of vibrating needle along the length direction of needle. Unit of the numbers for the horizontal axis in Fig. 6.2.3(a) is [m].

diameter is 0.83 mm in the section not tapered; the vibration frequency is 76.7 kHz, and horizontal vibration displacement in the vibration excitation section is 2.4 μm (0-peak); the needle is in the water film except the vibration excitation section. Fig. 6.2.3(a) shows the calculated sound pressure distribution around the needle, and Fig 6.2.3(b) shows the calculated sound pressure amplitude on the needle surface, in which z is the distance from the thick end of the needle (see Fig. 6.2.2). Comparing the positions of the droplet centers and sound pressure anti-nodes in Fig. 6.2.3(b), it is known that the droplet centers are at the sound pressure anti-nodes.

Based on the above stated experimental results and analyses, a physical model to explain the extraction and adsorption of droplets is proposed, as shown in Fig. 6.2.4. Only the sound field on one side of the needle is analyzed because of anti-symmetric property of the sound pressure about the needle. P2 and P3 are two adjacent nodal points of vibration on the needle, and P1 at $z = z_a$ is the anti-nodal point between them. Considering a rectangular area of water film S enclosed by the needle and boundary $L1$, $L2$ and $L3$, there is adhesive force acting on the water film by the needle, at the interface between the needle and water film. The force is caused by the interaction between oxygen atoms in the oxides of the needle surface and hydrogen atoms in the water molecules. Largeness of this force per unit length is assumed to be Fo. At boundary $L1$, there is cohesive force acting on the water film S by the water film outside S, which is caused by the H-bond. Largeness of this force per unit length is assumed to be F_h. When the needle has no vibration, F_h is larger than F_o. Hence, droplets cannot adsorb onto the needle when the needle is lifted. When the needle is in vibration, F_h may become less than F_o because of the negative cycle of sound pressure. In this case, droplets can adsorb onto the needle when the needle is lifted. The droplets lifted out of the water film may take the shape shown in Fig. 6.2.1(c) because of the intermolecular attractive force among molecules. Also, because the sound pressure is maximum in the vicinity to P1 at $z = z_a$, F_h is the minimum at $z = z_a$ and F_o - F_h is the maximum at $z = z_a$; because the sound pressure is zero in the vicinity to P2 and P3, F_o - F_h is the minimum in the vicinity to P2 and P3. For this reason, the centers of adsorbed droplets are at the anti-nodal points (P1) of needle vibration.

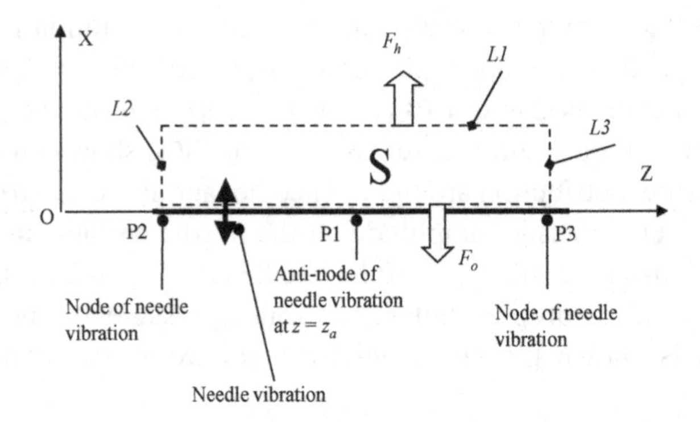

Fig. 6.2.4. Physical model for the extraction and adsorption of droplets.

6.2.3 *Characteristics and discussion*

Figure 6.2.5 shows the top and side views of adsorbed water droplets when the vibration displacement (peak-peak) at point P is 3.5 μm [Fig. 6.2.5(a)] and 1.8 μm [Fig. 6.2.5(b)], respectively, for the needle of 0.72 mm-diameter. The images of top view are taken from the

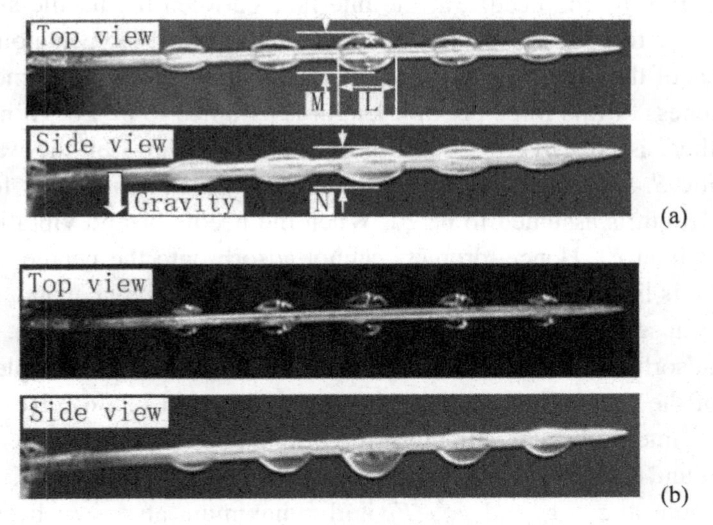

Fig. 6.2.5 The shape of water droplets adsorbed on the stainless needle of 0.72 mm diameter. The peak-peak vibration displacement at point *P* is 3.5 μm in (a) and 1.8 μm in (b). Point *P* is at the needle root (z = 7 mm).

gravitational direction, and the ones of side view are taken from the direction perpendicular to the page in Fig. 6.2.1(b). It is observed that adsorbed water droplets take on the shape of ellipsoid and are symmetric about the needle when the vibration displacement at point P is large

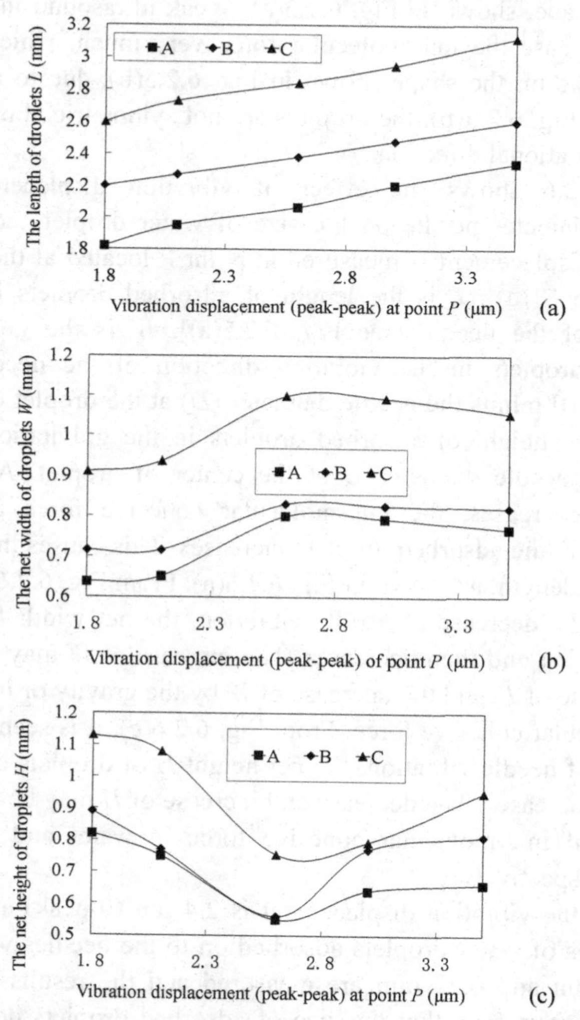

Fig. 6.2.6 Dependence of the size of adsorbed water droplets A, B and C on the vibration diaplacement of needle at point P. (a) The droplet length vs. vibration; (b) the net droplet width vs. vibration; (c) The net droplet height vs. vibration. The diameter of acoustic needle is 0.72 mm. Point P is at the root of needle (z = 7 mm) .

enough, and adsorbed water droplets are not symmetric about the needle when the vibration displacement at point *P* is small. Strong ultrasound in the droplets makes the intermolecular force among water molecules small, which causes the adsorbed droplets to take on a symmetric ellipsoid shape, shown in Fig. 6.2.5(a). Weak ultrasound in the droplets cannot decrease the intermolecular force very much, which makes the droplets take on the shape shown in Fig. 6.2.5(b), due to the action of gravity. In Fig. 6.2.5(b), the droplets are not symmetric about the needle in the gravitational direction.

Fig. 6.2.6 shows the effect of vibration displacement of the 0.72 mm-diameter needle on the size of water droplets adsorbed. The vibration displacement is measured at point P located at the needle root [see Fig. 6.2.1(a)]. *L* is the length of adsorbed droplets in the length direction of the needle [see Fig. 6.2.5(a)], *W* is the gross width of adsorbed droplets in the vibration direction of the needle (*M*) [see Fig. 6.2.5(a)] minus the needle diameter (*D*) at the droplet center, and *H* is the gross height of adsorbed droplets in the gravitational direction minus the needle diameter *D* at the center of droplet. As the needle vibration decreases, the intermolecular cohesive force among water molecules in the adsorbed droplets increases. This causes the decrease in the droplet length, as shown in Fig. 6.2.6(a). From Fig. 6.2.6(b), it is seen that with the decrease of needle vibration, the net width *W* of droplets increases first and then decreases. The increase of *W* may be caused by the decrease of *L*, and the decrease of *W* by the gravity or increase of the intermolecular cohesive force. From Fig. 6.2.6(c), it is seen that with the decrease of needle vibration, the net height *H* of droplets decreases first and then increases. The decrease and increase of *H* may be caused by the increase of intermolecular cohesive force in water and the effect of gravity, respectively.

When the vibration displacement is 2.4 μm (0-peak) at point *P*, the dimensions of water droplets adsorbed on to the needles with diameters of 0.83 mm and 0.72 mm are measured and the results are shown in Fig. 6.2.7. It is seen that the size of adsorbed droplets depends on the needle diameter, and the needle with larger diameter adsorbs larger droplets.

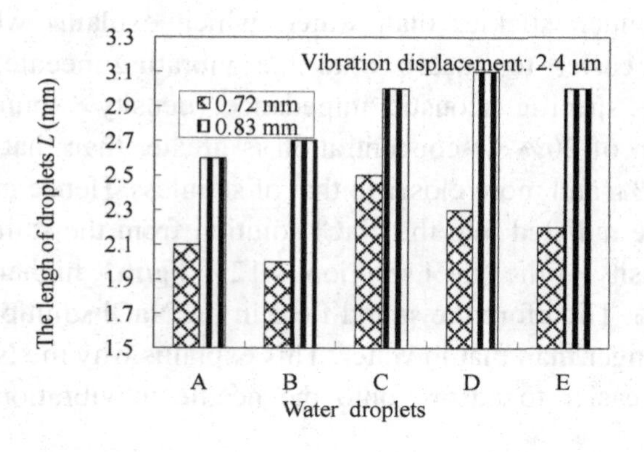

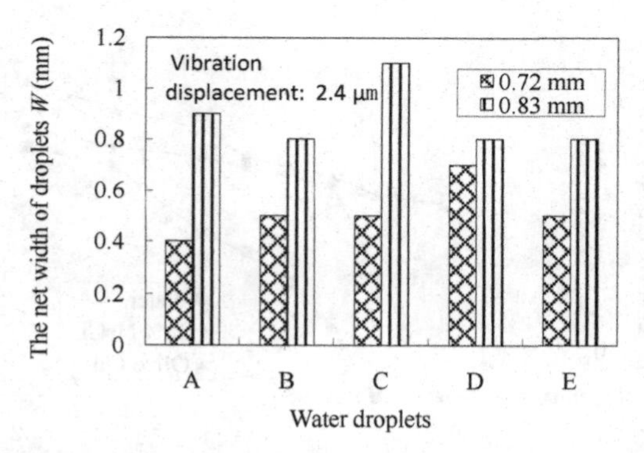

FIG. 6.2.7 The effect of needle diameter on the size of adsorbed droplets. (a) The length of droplets; (b) the net width of droplets. The vibration displacement is 2.4 μm (peak-peak) at point P, where is at the root of needle (z = 7 mm).

Olive oil and NaCl solution of 20% wt concentration are also used in the experiment. It is observed that their droplets can adsorb onto the needle in vibration, with the pattern shown in Fig. 6.2.1(c). Fig. 6.2.8 shows the measured length of droplet A versus vibration displacement at point P for the olive oil, NaCl solution and water. It is seen that the olive oil and NaCl solution droplets are easier to adsorb onto the needle in vibration than water droplets. The density (912 kg/m^3) and sound speed (1431 m/s) of olive oil are quit close to those of water. Hence the acoustic energy transmission coefficient from stainless to olive oil is close to that from stainless to water. However, the viscosity of olive oil (0.084 Pas) is much larger than that of water (0.001 Pas). This means that olive oil is much stickier than water, which explains why olive oil droplets are easier to adsorb onto the vibrating needle than water droplets. The specific acoustic impedance (density × sound speed) of NaCl solution of 20% wt concentration is greater than that of water by more than 25%, and more closer to that of stainless. Hence more acoustic energy can be radiated into the NaCl solution from the stainless needle. Also, the density of the NaCl solution is 1250 kg/m3, higher than that of water by 25%. Therefore the sound field in the NaCl solution around the needle is stronger than that in water. This explains why the NaCl solution droplets are easier to adsorb onto the needle in vibration than water droplets.

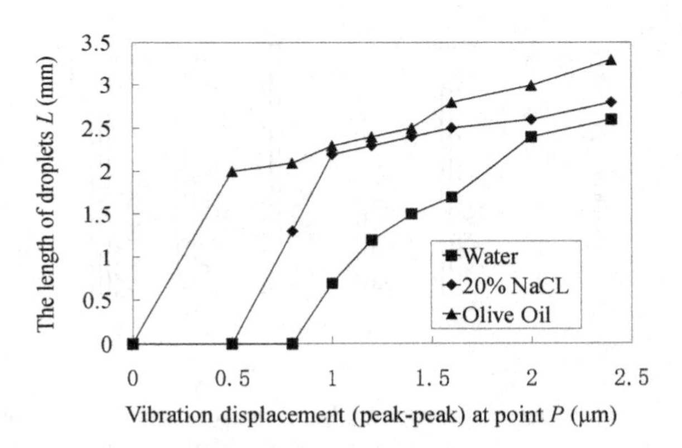

Fig. 6.2.8 The length of adsorbed droplet A versus vibration diaplacement for NaCl solution of 20% wt concentration, olive oil and water. The diameter of acoustic needle is 0.83 mm.

Fig. 6.2.9 shows the measured length of adsorbed droplets of pure water and saline solutions with 10%, 20% and 26% wt concentrations versus vibration displacement at point P. The needle diameter is 0.83 mm in the section not tapered. It is seen that the minimum vibration displacement to adsorb droplets decreases as the NaCl concentration increases. This is because more acoustic energy can be radiated into the NaCl solution from the needle as the concentration of NaCl increases, as explained in the discussion for Fig. 6.2.8.

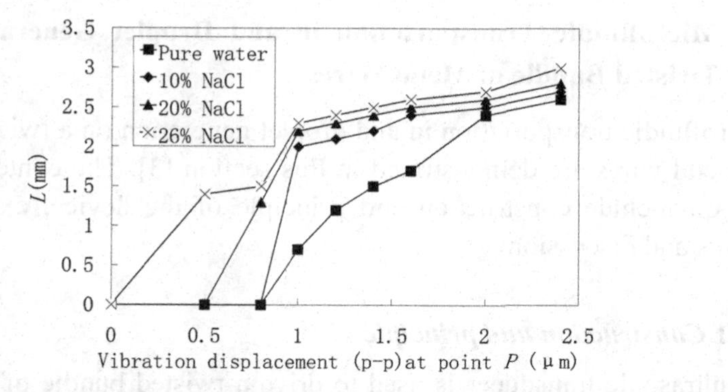

Fig. 6.2.9. The length of adsorbed droplets versus vibration displacement at point P for saline solutions of different concentrations.

Fig. 6.2.10 is a photograph showing the adsorbed olive oil droplets. The distribution of olive oil droplets is similar to that of water droplets except a small oil droplet at the needle tip. It is observed that a small water droplet adsorbed onto the needle tip is very quickly atomized after it is lifted from water film. Thus the small oil droplet at the needle tip may result from larger viscosity of olive oil. Due to larger viscosity, it is more difficult for the olive oil droplet to atomiz.

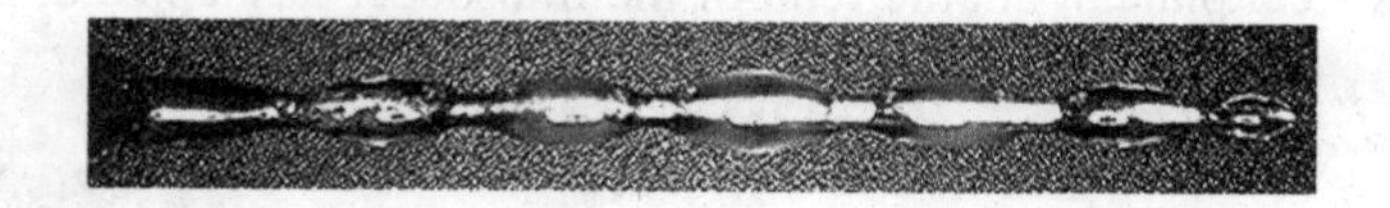

Fig. 6.2.10. Image of olive oil droplets adsorbed onto the vibrating needle of 0.83 mm diameter.

6.2.4 *Summary*

This section gives a method which utilizes the anti-nodes of a stainless needle in flexural vibration to adsorb liquid droplets. The adsorption capability depends on the needle vibration and liquid properties such as viscosity, specific acoustic impedance, and density, and the needle diameter. The method proposed in this paper can be used to extract droplets from liquid without a chamber.

6.3 Microfluidic Transportation in and Droplet Generation on a Twisted Bundle of Metal Wires

Microfluidic transportation in and droplet generation on a twisted bundle of metal wires are demonstrated in this section [3]. The contents of this section include construction and principle of the device, experimental results and discussion.

6.3.1 *Construction and principle*

An ultrasonic transducer is used to drive a twisted bundle of thin metal wires mechanically (Fig. 6.3.1). The bundle of metal wires is clamped by two stainless steel plates at one end of the ultrasonic transducer by a bolt structure. Ten copper wires with a diameter of 199 μm each form the bundle, and the bundle length is 38 mm beyond the stainless steel plates. The ultrasonic transducer has a sandwich structure with a multilayer piezoelectric vibrator at the center, which is formed by four piezoelectric rings (C203, Fuji Ceramics, Japan). The outer diameter, inner diameter and thickness of each piezoelectric ring are 20mm, 12 mm and 2.4mm, respectively. The two stainless steel plates clamping the bundle of metal wires have a size of 20 mm × 20 mm × 2.5 mm each and the stainless steel plate at another end of the transducer has a size of 20 mm

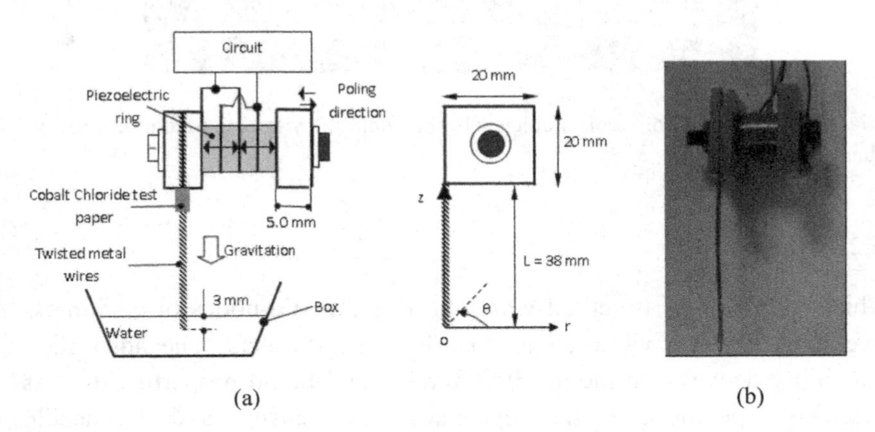

Fig. 6.3.1. An ultrasonic actuator with a twisted bundle of thin copper wires for microfluidic delivery. (a) Schematic diagram. (b) Photograph.

× 20 mm × 5 mm. A torque of 26 kgcm is applied to the bolt to tighten the whole structure. Resonance frequency of the whole system is about 66.35 kHz at low vibration. The bundle of metal wires is twisted properly to obtain some micro channels among the wires. The width of the micro channels formed among the metal wires can be controlled by the twist turns.

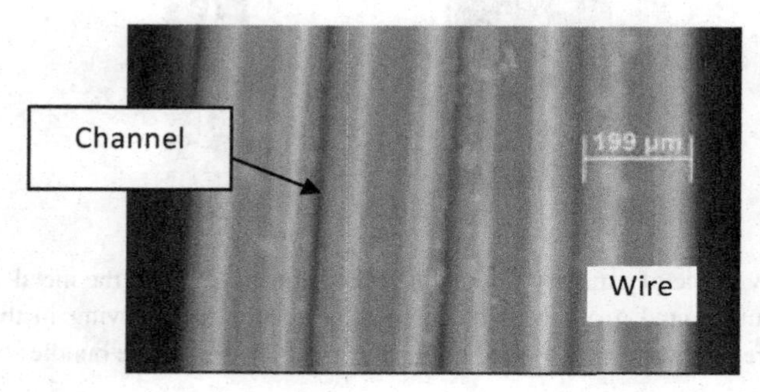

Fig. 6.3.2. Magnified image of the micro channels and copper wires. The image was taken by a microscope system (Olympus BX-51, Olympus, Japan).

In the experiments, the metal wire bundle was twisted 8 turns and the average width of the channels was about 12 μm (Fig. 6.3.2). The length of the metal wire bundle immersed in water was about 3 mm. When a driving voltage with a frequency close to the resonance frequency was applied to the ultrasonic transducer, the thickness mode vibration could be excited in the transducer. This resulted in an ultrasonic vibration in the thin metal wires. It was observed that water ascended along the metal wires when the driving voltage and its frequency were appropriate. To observe the traces and transportation of water, UV reactive fluorescent water and a high intensity ultraviolet lamp with a wavelength of 365 nm and power of 100 Watt (Blak-Ray B-100AP, UVP, Cambridge, UK) were used. Fig. 6.3.3(a) shows the traces of green fluorescent water (Big Water, Thermaltake, Singapore) ascending along the channels between the wires, and Fig. 6.3.3(b) the transportation of red fluorescent water, which is a 0.85 g/ml solution of red fluorescent dye (Model No: 00295-16, Cole-Parmer Instrument, USA). From Fig. 6.3.3(a), it is seen that the ascending water is in the gaps among the metal wires rather than on the whole surface of the metal wire bundle. From Fig. 6.3.3(b), it can be clearly seen that red fluorescent particles are transported to the gap between the two metal plates clamping the wires. When the metal wires had no vibration, the above phenomena could not be observed.

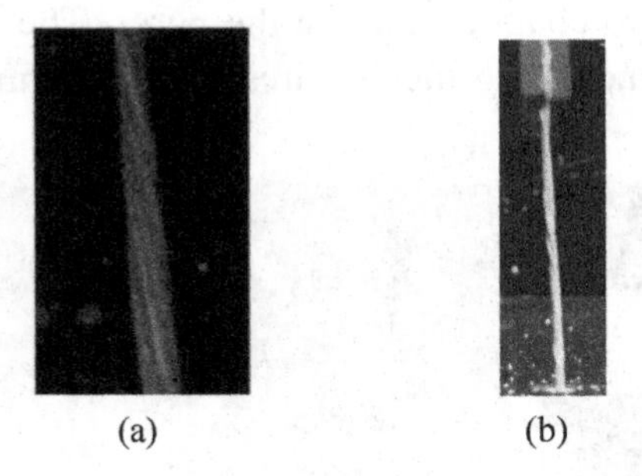

Fig. 6.3.3. Evidence of capillary flow in the micro channels within the metal wire bundle induced by ultrasound. (a) Traces of green fluorescent water moving in the channels. (b) Red fluorescent water transported to the upper part of metal wire bundle.

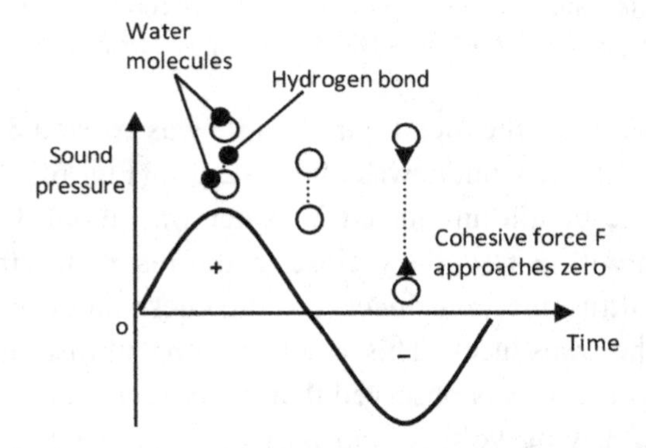

Fig. 6.3.4. A physical model to describe the effect of sound pressure on cohesive force between the molecules of liquid. The circles represent the molecules in liquid.

During the negative half circle of sound pressure, negative sound pressure increases the distance among the molecules in liquid greatly, which is also the reason for acoustic cavitation. In this case, the cohesive force among the molecules becomes small. During the positive half circle of sound pressure, positive sound pressure compresses the molecules in liquid. But this cannot increase the cohesive force too much because the molecules repel each other when they are too close. On the time average, ultrasound may make the cohesive force among the molecules of liquid weak; thus the adsorption force between the capillary tube and liquid may become larger than the cohesive force. The sound induced change of cohesive force among the molecules can be expressed by Fig. 6.3.4. Based on the model, ultrasound may enhance the capillary action in the gap.

6.3.2 *Results and discussion*

To measure the capillary speed of water, a piece of Cobalt Chloride test paper (Kaagat Ltd, England) was clipped round the upper part of the metal wires [Fig. 6.3.1(a)]. The test paper was blue when it was dry,

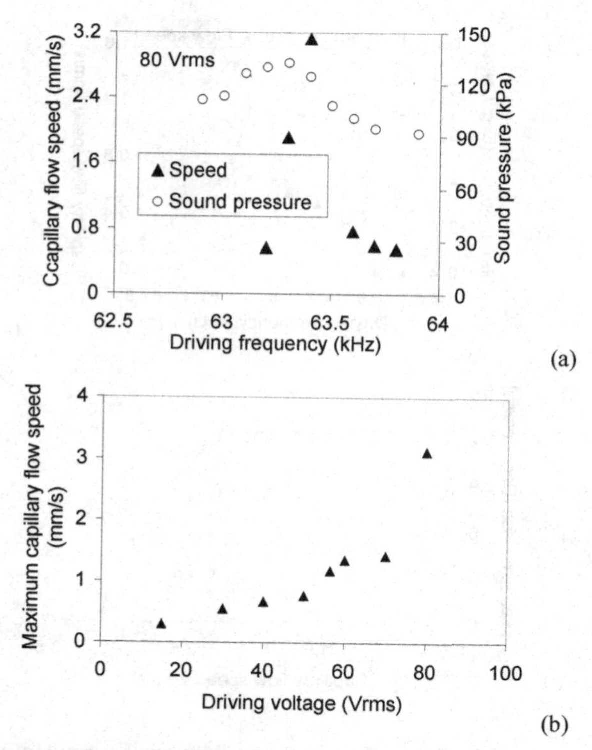

Fig. 6.3.5. Dependence of the capillary flow speed on driving frequency, sound pressure and driving voltage for water. (a) Effect of the driving frequency and sound pressure on the capillary flow speed. (b) Maximum capillary flow speed for driving frequency versus driving voltage.

and became pink when the water reached it through the channels. The test paper had to be heated by a hot air blower at about 50°C before use. The speed was obtained by measuring the time it took for the water to move from the water surface to the center of test paper. Thus it represented the average speed along the channels. After each measurement, the metal wires were dried by an air blower to ensure that the metal wire bundle used for the next measurement was dry. In the following experiments and discussion, unless otherwise specified, the number of copper wires used was ten; they were twisted 8 turns; the length of metal wires outside the metal plates was 38 mm; water was used as the liquid; the length of the metal wires in water was 3 mm.

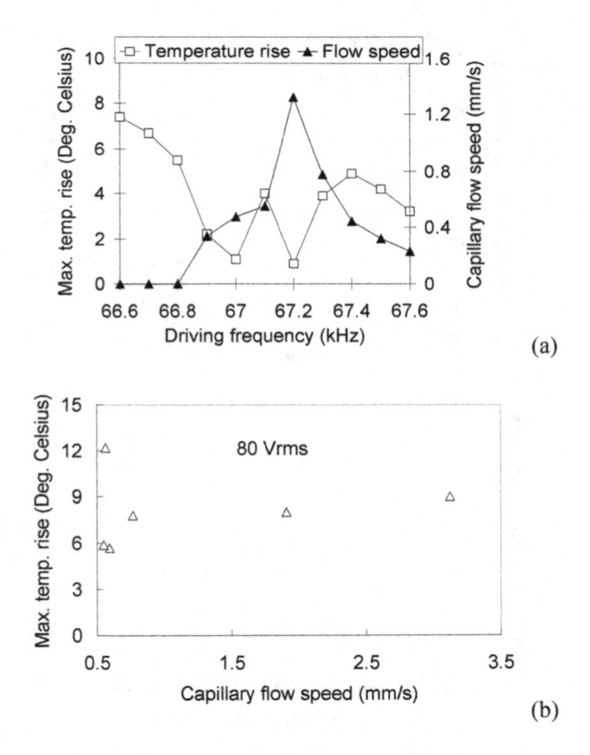

Fig. 6.3.6. Investigation of the effect of temperature on the capillary phenomenon for water. (a) Maximum temperature rise of transducer and capillary flow speed versus driving frequency. (b) Maximum temperature rise versus capillary flow speed.

The dependence of speed of the capillary flow in the micro channels on the driving frequency, sound pressure and driving voltage was measured, and the results are shown in Fig. 6.3.5. The sound pressure was measured by a 1 mm needle hydrophone (SN 945, Precision Acoustics, UK) at $r = 0.9$ mm, $\theta = \pi/2$, $z = 0$, which is perpendicular to the metal wires. The speed reaches a maximum at around the resonance point [Fig.6.3.5(a)]. This phenomenon can be explained by the strong sound pressure at the resonance point. Also, the maximum capillary flow speed with respect for driving frequency increases as the driving voltage increases [Fig. 6.3.5(b)]. In Fig. 6.3.5(b), operating frequency to get the maximum flow speed depends on the driving voltage, and is not a constant. As the driving voltage increases from 15 Vrms to 80 Vrms, the operating frequency decreases from 67.2 kHz to 63.4 kHz. This phenomenon can be well observed in many ultrasonic actuators, which is due to the heat generation.

It is known that the temperature of liquid affects the capillary speed in a capillary phenomenon without ultrasound. As the temperature increases, the distance between the molecules in liquid increases; thus the adhesive force among molecules decreases. This may enhance the capillary action. To investigate the effect of temperature in the experiments, temperature of the ultrasonic transducer including the metal wire bundle is monitored during operation. It was found that the middle of the multilayer piezoelectric vibrator had the highest temperature. Fig. 6.3.6(a) shows the relationship among the highest temperature rise and capillary flow speed. The relationship was measured by scanning the driving frequency from 66.6 kHz to 67.6 kHz at a driving voltage of 56.42 Vrms. An infrared thermal meter (Keyence IT2-50, Keyence Corp.) was used in the measurement of temperature. Water temperature in the micro channels must be proportional to the highest temperature at the piezoelectric vibrator because the heat conduction is linear due to the relative low temperature. Hence, from Fig. 6.3.6(a), it can be deduced that the capillary flow speed does not increase as the temperature increases. Actually it reaches the maximum when the temperature rise is the lowest. Therefore the temperature rise does not contribute to the capillary phenomenon in the experiments due to its small value. Fig. 6.3.6(b) shows the temperature rise versus capillary flow speed corresponding to Fig. 6.3.5(a). It supports the conclusion deduced from Fig. 6.3.6 (a).

To measure the transported mass, the change of mass of the box which contains the water for test [Fig. 6.3.1(a)] was measured by an electronic balance with a resolution of 0.01 g (TP2KS, Precision Plus, Singapore). To eliminate the influence of water evaporation from the water surface, a lid with a small hole was used to cover the box. The

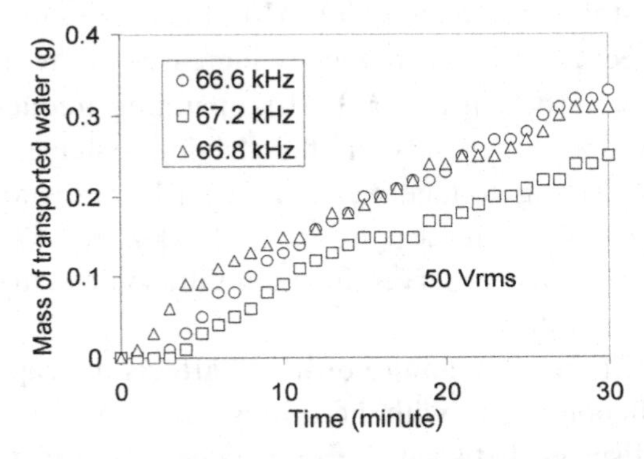

Fig. 6.3.7. Mass of transported water versus transportation time at a driving voltage of 50 V_{rms} and different driving frequencies.

diameter of the lid and hole was 130 mm and 8 mm, respectively. The metal wire bundle was inserted into the water through the hole to suck water. It was experimentally confirmed that the reading of electronic balance had no change when the driving voltage was zero and the lid was used. The test paper was used in the upper part of metal wires to monitor the capillary flow and suck the ascending water away from the bundle of metal wires. Without the paper, all of the suck up water will be in the bundle of metal wires and thus the reading of the electronic balance will have no change. When the transported water was sucked away by the test paper, the reading of electric balance had a change because the clip holding the test paper was hung by a wire and its mass was not included in the reading of electronic balance. So the change of the reading of electronic balance could be approximately viewed as the mass transported by the metal wires. Based on this, the mass of water transported by the bundle of metal wires versus transportation time was measured and the result is shown in Fig. 6.3.7. The mass of transported water per unit time can be estimated from Fig. 6.3.7, which is the slope of trend lines of the points. At the operating frequency of 66.6 kHz and driving voltage of 50 Vrms, it is about 2×10^{-4} g/s. As a whole, this slope changes little for a given voltage and operating frequency. Hence, the ultrasound induced capillary flow in the micro channels is quite stable.

To investigate the capillary flow of a liquid with larger viscosity than water, vegetable oil with a density of 994 kg/m^3 and viscosity of 0.035 Pa.s was used. Fig. 6.3.8 shows the capillary flow speed versus driving frequency at a driving voltage of 80 Vrms. It is seen that the maximum capillary speed is 0.56 mm/s for vegetable oil at 80 Vrms, which is much smaller than that for water (3.2 mm/s). Due to the larger viscosity of vegetable oil, flow resistance is larger and sound pressure is weaker in the channels.

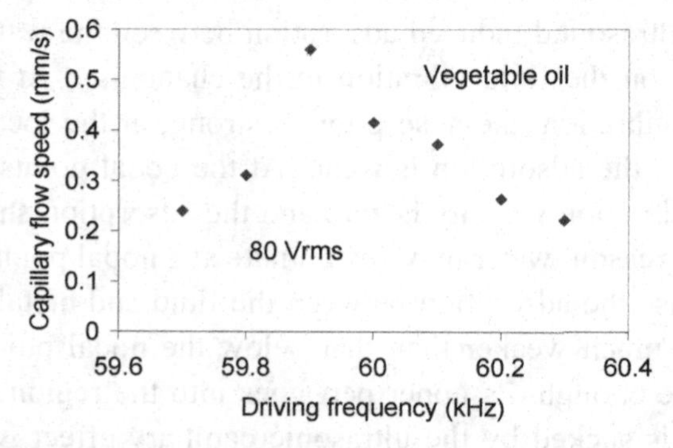

Fig. 6.3.8. Capillary flow speed versus driving frequency for vegetable oil at a driving voltage of 80 Vrms.

During the experiments, it was also observed that the water ascending within the metal wire bundle could form some droplets on the surface of the metal wires if the vibration was strong enough; the size of the droplets gradually increased to the maximum and then diminished to zero, which could last for up to 20 seconds. The photograph shown in Fig. 6.3.9 was taken by using the green fluorescent water and UV light when the metal wires were in vibration. It shows two droplets formed on the metal wires. After switching off the driving voltage, the droplets could remain on the metal wires. These phenomena provide a method to

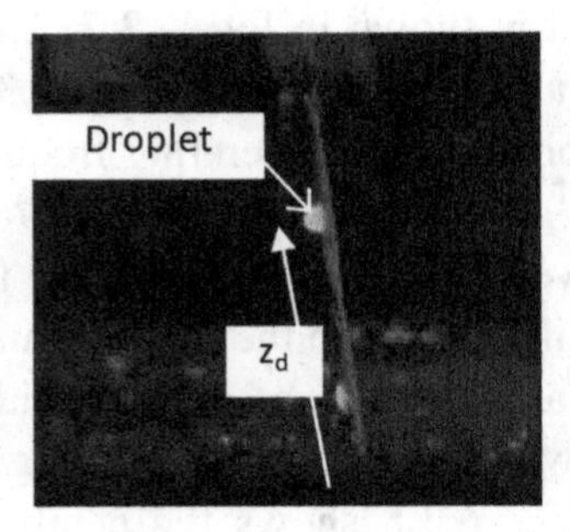

Fig. 6.3.9. Water droplets on the surface of the metal wire bundle. Green fluorescent water and UV light were used in taking the photograph.

extract the liquid sucked up by the metal wire bundle. The radii of droplets can be calculated from the photo. For the upper droplet, it is about 1.7 mm. These phenomena may be explained qualitatively as follows. The ultrasound induced adsorption between the water and metal wires depends on the fluid vibration in the channels. At the locations with a strong vibration, the adsorption is strong; at the locations with a weak vibration, the adsorption is weak. At the nodal points of vibration velocity, the vibration velocity is zero and the adsorption should be very weak. For this reason, water may accumulate at a nodal point of vibration velocity because the adsorption between the fluid and metal wires at the nodal point is much weaker than that below the nodal point. When the droplet is large enough, its upper part goes into the region with enough vibration and is sucked by the ultrasonic capillary effect, which causes the decrease of the water droplet size.

The effects of the driving voltage and frequency on the location of the droplets were investigated experimentally, and the results are shown in Figs. 6.3.10(a) and 6.3.10(b). It is seen that the location is independent of the driving voltage and frequency near the resonance point, and there is always a droplet at $z_d = 14$ mm, where z_d is the distance from the tip, as shown in Fig. 6.3.9. The sound speed c in copper wires is 3700 m/s; the resonance frequency f_r is approximately 66.4 kHz. Using $\lambda = c/f_r$, the wavelength in copper wires is calculated to be 55.7 mm. From the vibration analyses of a beam or a string, it is known that the vibration of the metal wires is maximal at the tip. Thus there must be a nodal point of vibration velocity at $z_d = \lambda/4$, which is calculated to be 13.9 mm. This nodal point of vibration results in a nodal point of sound pressure at $z_d =$ 13.9 mm. This explains why there is always a droplet around $z_d = 14$ mm, which is shown by Fig. 6.3.10. A bundle of metal wires with a length of 90 mm beyond the transducer was used. The resonance frequency f_r of the transducer was about 56.8 kHz. Fig. 6.3.11 shows the measured location of one of its droplets. It is seen that the droplet is also near $z_d \approx 0.25\lambda$. From the above results, it can be derived that the location of one of the droplets on metal wires may be predicted by $z_d = c/(4f)$, where c is the sound speed in the metal wires and f the driving frequency.

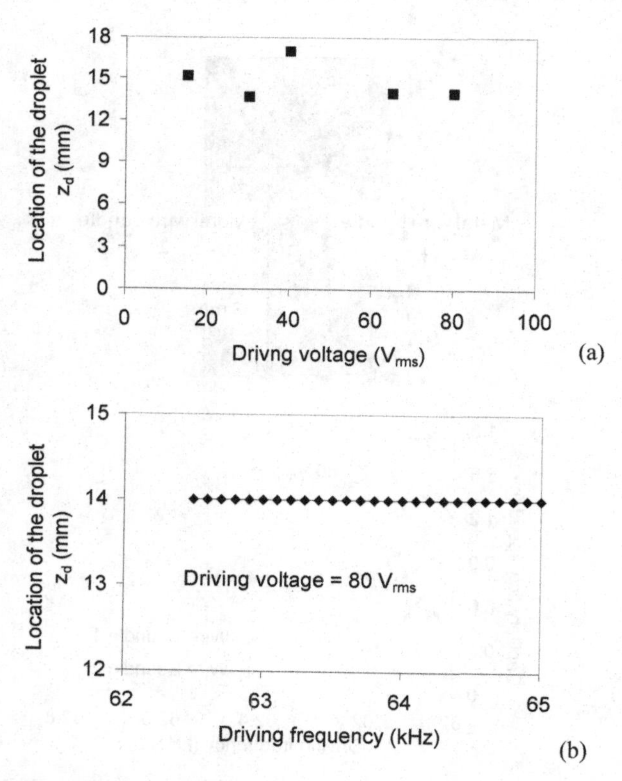

Fig. 6.3.10. Location of one of the water droplets for a transducer with L = 38 mm. (a) Location versus driving voltage near the state of maximum capillary flow speed. (b) Location versus driving frequency at a driving voltage of 80 Vrms.

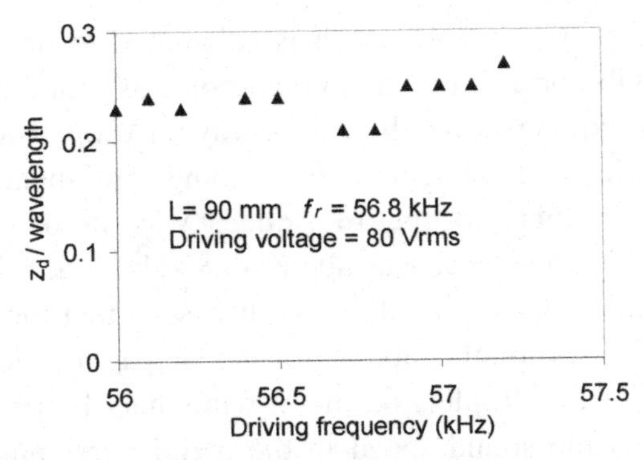

Fig. 6.3.11. Location of one of the water droplets versus driving frequency for a transducer with L = 90 mm, resonance frequency of 56.8 kHz, and driving voltage of 80 Vrms.

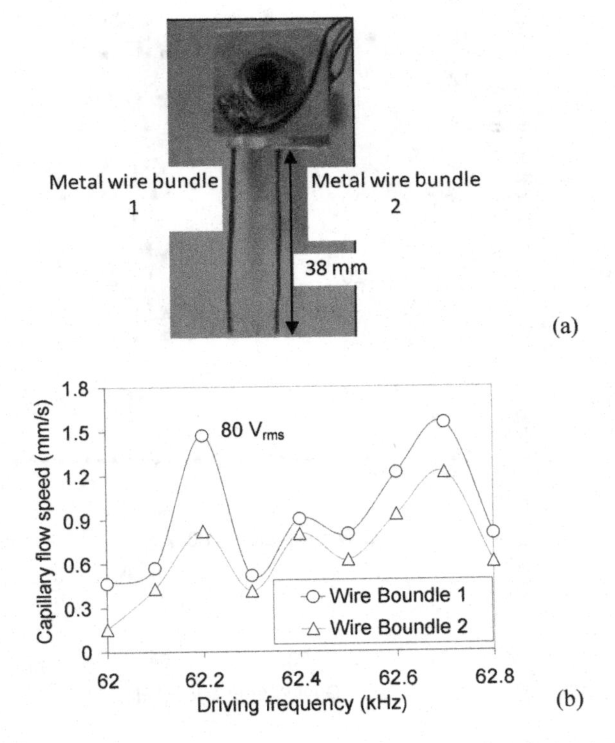

Fig. 6.3.12. Ultrasound induced capillary flow in two twisted metal wire bundles mechanically excited by the transducer simultaneously. (a) A transducer with two parallel bundles of metal wires. (b) Capillary flow speed versus driving frequency in the two bundles of metal wires at a driving voltage of 80 Vrms.

The droplets could always be obtained when a dry metal wire bundle was used. However, sometimes droplets could not be formed when a wet metal wire bundle was used. The latter phenomenon was because there was an adsorption between the fluid and metal wires above a nodal point of vibration velocity when the metal wires were wet. When the vegetable oil was used, no droplet was observed during the experiments. This could be due to a high viscosity of the oil.

To investigate the possibility of using an array of metal wire bundles, two metal wire bundles as shown in Fig. 6.3.12(a) were used and the capillary flow speed in these two metal wire bundles was measured. Each metal wire bundle had ten wires, and was twisted 8 turns. The result shown in Fig. 6.3.12(b) indicates that it is possible to assemble

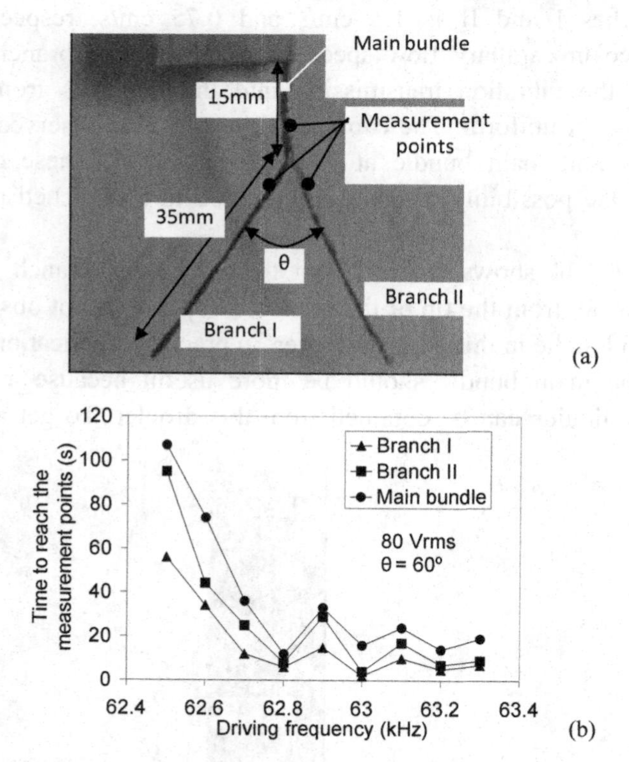

Fig. 6.3.13. Ultrasound induced capillary flow in a branched metal wire bundle. (a) Photograph of the branched metal wire bundle. (b) Time to reach the measurement points versus driving frequency at a driving voltage of 80 Vrms and θ = 60°.

multiple metal wire bundles to suck up water simultaneously. Due to the non-uniformity of vibration on the edge of the metal plates, the capillary flow speed has a difference in the two metal wire bundles.

To explore the potential of applying the method to a branched microfluidic system, a branched metal wire bundle shown in Fig. 6.3.13(a) was tested. It was found that water could reach the main bundle via the two branches. Fig. 6.3.13(b) shows the time for water to arrive at the measurement points versus driving frequency at $\theta = 60°$ and 80 Vrms driving voltage. Branches I and II had 10 metal wires each, and the main channel had 20 metal wires. The two branches were twisted 8 turns. The distance from the junction to the measurement points was 5 mm. From Fig. 6.3.13(b), it is known that the maximum capillary speed in branches I and II is 1.5 cm/s and 0.75 cm/s, respectively. The difference in capillary flow speed between the two branches may be because the vibration transmission into the branches from the main bundle is not uniform. The capillary flow was also observed in the two branches and main bundle at $\theta = 30°$ and 120°. These experiments indicate the possibility to apply the method to a branched microfluidic system.

Fig. 6.3.14 shows a droplet on the surface of branch I, which is 14 mm away from the tip of the branch. Droplet was not observed along the main bundle in this case. However, in practical applications, a droplet along the main bundle should be more useful because a mixture of different liquids can be obtained from this droplet. To get a droplet on

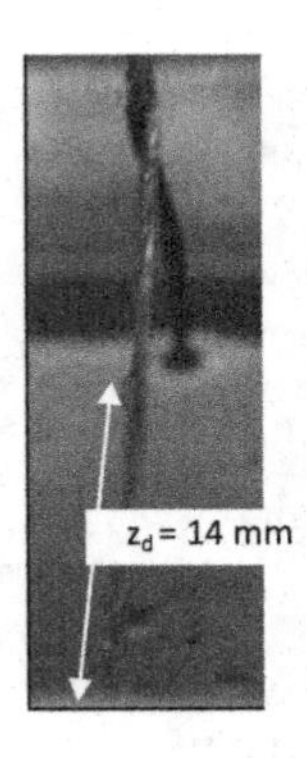

Fig. 6.3.14. Water droplet on the surface of branch I.

the main bundle, the length ratio of the main bundle to branch was changed from 15:35 to 25:25. The transducer with the changed length ratio was tested with an immersed length of 12 mm at a driving frequency of 75 kHz, which is near another resonant peak. A droplet was observed 5 mm away from the junction on the surface of the main bundle (Fig. 6.3.15). Due to the existence of the junction, it becomes complicated to predict the location of droplets.

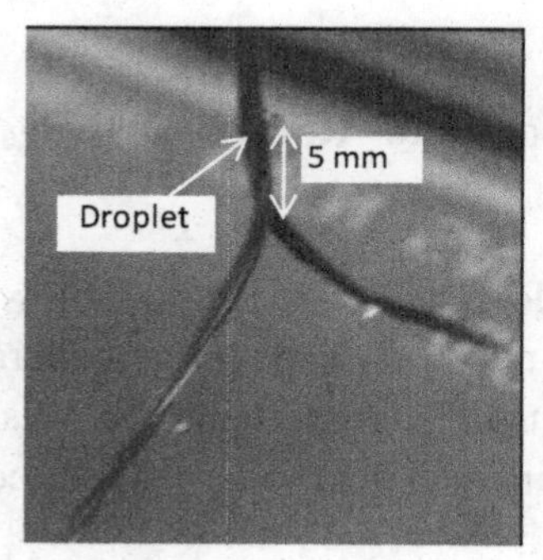

Fig. 6.3.15. Water droplet on the surface of the main bundle with a changed length ratio.

6.3.3 *Summary*

This section describes the ultrasonic actuator that uses ultrasound induced capillary action in a twisted bundle of metal wires to transport tiny quantity of liquid. To make the bundle of metal wires vibrate ultrasonically, it is mechanically excited by a sandwich type ultrasonic transducer. Liquid is transported in the micro channels among the twisted metal wires. The transportation of water through the bundle of metal wires and its dependence on ultrasonic vibration are experimentally confirmed, and a physical model is proposed to explain the principle.

The capillary flow speed versus driving frequency is measured for water and vegetable oil. The speed reaches a maximum near the resonance point. The vegetable oil which has a much higher viscosity than and close density to water, has a lower capillary speed than water. Transported water volume per unit time is measured experimentally. According the measurement, it is about 0.2 mg/s at a driving voltage of 50 Vrms and frequency of 66.6 kHz. The capillary flow is also investigated when a branched bundle of metal wires is used. The results show that the method may be used in branched multi-channel microfluidic systems. Furthermore, water droplets can be obtained on the surface of the metal wire bundles, which provides a method of extracting the liquid transported in the micro channels. Experimental analysis shows that they are caused by the vibration nodal points along the metal wires. The maximum capillary speed in the experiments is about 15 mm/s.

6.4 Droplet Generation and Rotary Driving with an Ultrasonic Hollow Needle

In this section, it is demonstrated that a hollow needle mechanically driven by an acoustic needle is capable to generate and rotate a small water droplet at its tip. The contents of this section include the experiment setup, principle analyses, measured characteristics and discussion.

6.4.1 *Experimentation and operating mechanism*

Fig. 6.4.1 shows the experimental setup for the generation and rotation of droplet. A hypodermic needle with reservoir is filled with water, and excited to vibrate ultrasonically in the z-direction by an acoustic needle, as shown in Fig.6.4.1(a). The hypodermic needle (BD Precision Glide 30G) is made of stainless steel and has a blunt tip. It has a length of 10.5 mm, inner diameter of 172 μm and outer diameter of 311 μm as shown in Fig.6.4.1(b). The acoustic needle is driven by an ultrasonic transducer which is clamped onto the arm of an xyz stage. The tip of the acoustic needle is located at position P ($x = 0$, $y = 6$ mm, $z = 0$), in solid contact with the hypodermic needle. Fig. 6.4.1(c) shows the ultrasonic transducer used in the experiments. Four piezoelectric ceramic discs are pressed against each other between two 20 mm × 20 mm × 2.3 mm stainless steel plates with a high-tensile steel screw running through the centre of the transducer, clamping the ceramic discs together. An acoustic needle is welded onto the front side of the stainless plate. It has a length of 34 mm from the edge of the stainless steel plate, and diameter

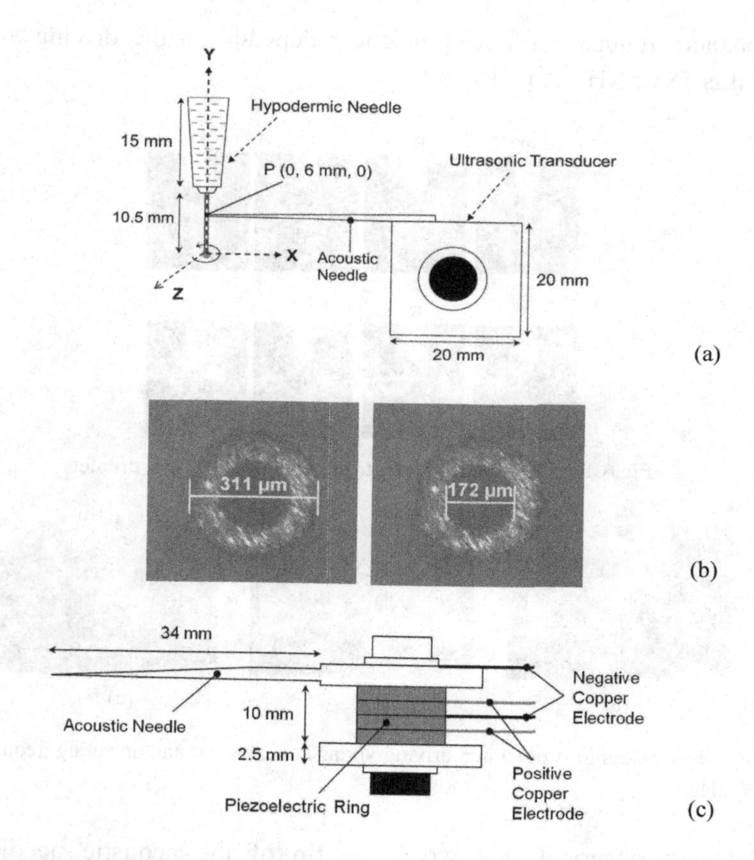

Fig.6.4.1. Experimental setup. (a) A hypodermic needle driven by ultrasonic transducer; (b) cross section of the hypodermic needle; (c) structure of the ultrasonic transducer.

of 0.8 mm which gradually decreases down to its tip. The piezoelectric rings have an inner diameter of 6 mm, outer diameter of 12 mm and thickness of 1.2 mm. The electromechanical quality factor Q_m, piezoelectric coefficient d_{33}, and relative dielectric constant $\varepsilon_{33}^{T} / \varepsilon_0$ of the piezoelectric rings are 2000, 325×10^{-12} m/v and 1450, respectively. To excite the first order thickness mode vibration in the transducer efficiently, any two adjacent rings have opposite poling direction. To drive the transducer, an AC driving voltage with a frequency near the resonance of the transducer is applied to the copper electrodes. The resonance frequency of the transducer depends on the driving voltage and it is 58.05 kHz at 15 Vrms.

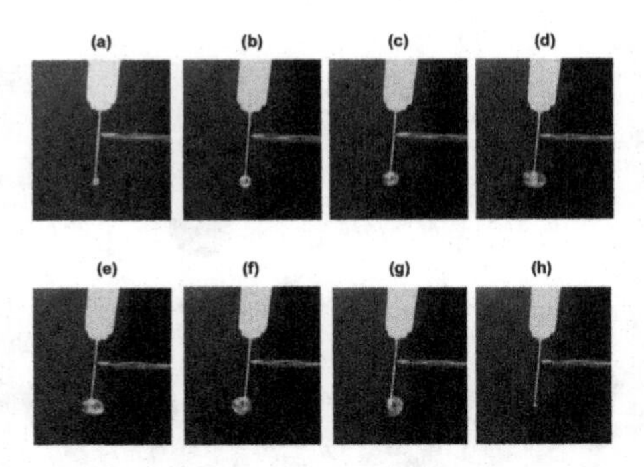

Fig.6.4.2. A series of images to show the growth of a droplet.

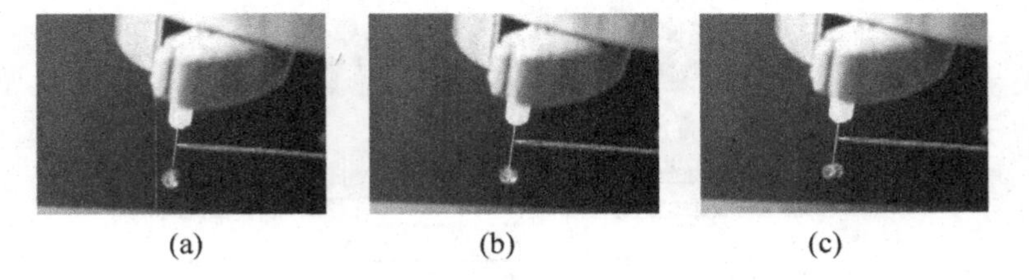

(a) (b) (c)

Fig. 6.4.3. Rotation of droplet at a driving voltage of 20 Vrms and operating frequency of 57.90 kHz.

It was observed that when the tip of the acoustic needle was perpendicularly in solid contact with the hypodermic needle at position *P* ($x = 0$, $y = 6$ mm, $z = 0$), a single water droplet began to form at the tip of the hypodermic needle if the ultrasonic transducer was driven by a voltage from 10 Vrms to 30 Vrms at resonance. When the driving voltage was less than 10 Vrms or greater than 30 Vrms, it was difficult to form a droplet because of a weak ultrasound in the hypodermic needle or atomization of the droplet. Note that no droplet was observed when there was no contact between the acoustic needle and hypodermic needle, and when ultrasonic vibration was not excited. The size of the droplet increases with the driving time. Upon reaching a diameter of 0.15 cm, the droplet began to rotate around the tip of the hypodermic needle. Fig. 6.4.2 shows the growth of a droplet at a driving voltage of 20 Vrms and frequency of 57.8 kHz. The time interval between two adjacent frames is approximately 4 seconds. From images (a) to (g), the droplet diameter increases with the increase of driving time. When the droplet diameter reached some size, the droplet falls off, and a new droplet starts to grow again [images (g) & (h)]. Fig.6.4.3 illustrates the rotation of the water droplet in anti-clockwise direction (viewing from the top) around the hypodermic needle tip at a driving voltage of 20 Vrms and operating frequency of 57.90 kHz. The time interval between two frames is approximately 0.033 second as the video was recorded using a digital camera with 30 frames per second. When the driving voltage at resonance is below 15 Vrms approximately, the droplet oscillates around the tip and the rotation can only be observed when the driving voltage is from 15 Vrms to 30 Vrms.

Due to vibration in the hypodermic needle, the water contained in the hypodermic needle also ultrasonically vibrates. This may decrease the surface tension of the water contained in the hypodermic needle. During the negative half circle of sound pressure, negative sound pressure increases the distance among the molecules of water, which is also the reason for acoustic cavitation effect. In this case, the cohesive force among the molecules becomes weak. During the positive half circle of sound pressure, positive sound pressure compresses the molecules of water. But this cannot increase the cohesive force too much because the molecules repel each other when they are too close. On the time average, ultrasound may make the cohesive force among the molecules of liquid weak. The decrease of water surface tension in the hypodermic needle causes the water in the hypodermic needle to flow down, and the accumulated water forms a droplet due to surface tension.

The vibrating hypodermic needle and the droplet construct an ultrasonic system very similar with the ultrasonic motor driven by fluid directly [5]. So, the possible reasons for the rotation of droplet includes a travelling wave along the circumferential direction of the hypodermic needle [5], and the non-uniformity of sound field in the droplet along the circumferential direction [6]-[7]. When the droplet is small, due to the relative large effect of flow resistance of the hypodermic needle surface, no acoustic streaming can be generated, as observed in our experiments.

The fall of droplet is because the surface cohesive force between the droplet and needle tip is overcome by the gravitation.

6.4.2 *Characteristics and discussion*

To avoid the atomization of droplet and obtain enough driving force for rotation, the vibration excitation point should be a little bit shifted from a nodal point of vibration velocity on the hypodermic needle. Based on this theoretical estimation, a location on the hypodermic needle 6 mm away from the tip is chosen as the vibration excitation point.

The change of droplet diameter with time at 20 Vrms and 57.95 kHz was measured and the result is illustrated in Fig.6.4.4. From Fig.6.4.4, it is seen that the speed of growth in the droplet is particularly faster in the beginning and slows down as it reaches its maximum diameter. This experimental phenomenon probably results from the vibration decrease of hypodermic needle, which is caused by the increase of droplet size. The droplet falls off 18 s (lifetime of droplet) after the droplet starts to appear.

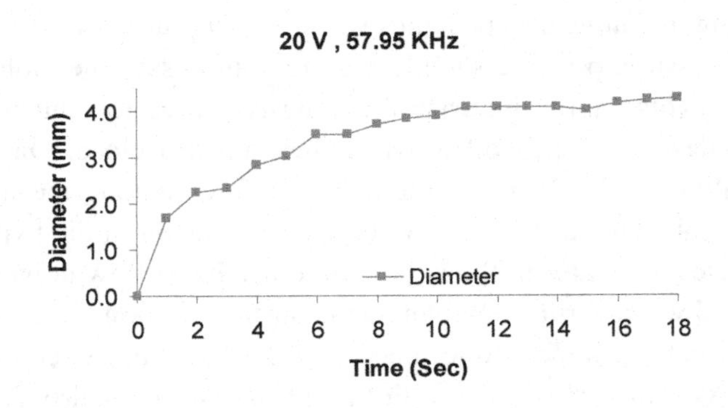

Fig. 6.4.4. Growth and maximum size of the droplet.

Fig.6.4.5 illustrates the maximum diameter of droplet versus operating frequency at 20 Vrms and 25 Vrms. The resonance frequency of the transducer at 20 Vrms and 25 Vrms is 57.95 kHz and 57.8 kHz, respectively. From this figure, it is seen that the diameter of the droplet reaches maximum at resonance. At resonance, the vibration of the hypodermic needle is maximal. Thus the experimental phenomenon indicates the ultrasonic wetting effect [8] between the outer surface of hypodermic needle and droplet is involved in sustaining the droplet.

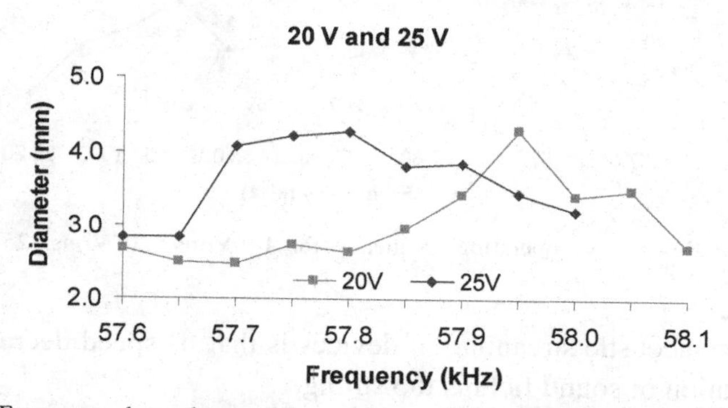

Fig. 6.4.5. Frequency dependence of the maximum diameter at 20 Vrms and 25 Vrms.

Fig.6.4.6 illustrates the number of revolutions per second versus operating frequency for 4 different voltages of 15 Vrms, 20 Vrms, 25 Vrms and 30 Vrms. The number of revolutions per second is measured by reducing the playback speed of each recorded video for the various frequencies and voltages. A peak in the number of revolutions per second is seen at each voltage level. Monitoring the phase difference between the input current and voltage, and the amplitude of input current, it is known that the frequency at which the peak of revolution speed occurs is the resonance frequency at the corresponding driving voltage. The peak resonance frequency is shifted from around 58.05 kHz at 15 Vrms, 57.95 kHz at 20 Vrms down to 57.80 kHz at 25 Vrms and subsequently 57.70 kHz at 30 Vrms. This shift is mainly due to the prolonged operation of the ultrasonic transducer, causing operating temperature to increase. Highest number of revolutions with an average of 16.7 revolutions per second (1002 rpm) and the widest frequency range which can produce revolutions are obtained at 25 Vrms. It is also seen that when the driving voltage is too large, the number of revolutions per second at the resonance decreases. This phenomenon supports the mechanism of droplet rotation proposed in Section 6.4.1. According to the work on actuators utilizing acoustic streaming [5-7, 9], it is known that a basic

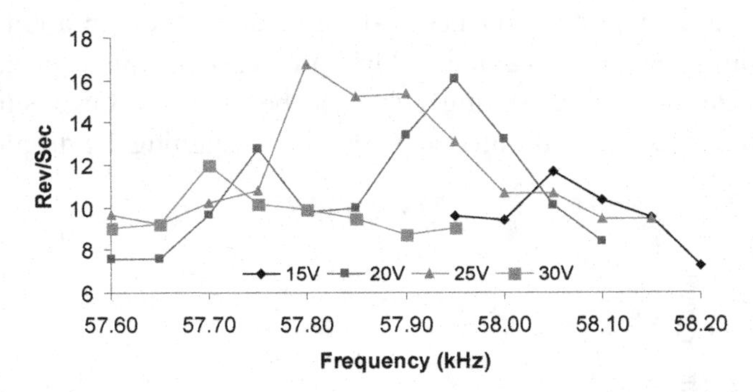

Fig. 6.4.6. Rev/Sec vs operating frequency for 15 Vrms, 20 Vrms, 25 Vrms, and 30 Vrms.

feature of acoustic streaming in devices is that its speed decreases when the vibration of sound field is too strong.

Lifetime of the droplet (time to generate and dispense one droplet) was also measured. Fig.6.4.7 illustrates the lifetime versus frequency for 15 Vrms, 20 Vrms, 25 Vrms and 30 Vrms. From the figure, it is seen that the lifetime is minimal at the resonance frequency. The speed of accumulation of water around the tip increases as the ultrasonic vibration of hypodermic needle becomes strong. So the speed of accumulation is maximal at resonance. Also, the average cohesive force among water molecules is the minimum at resonance. For these two reasons, the lifetime of droplet is minimal at the resonance frequency.

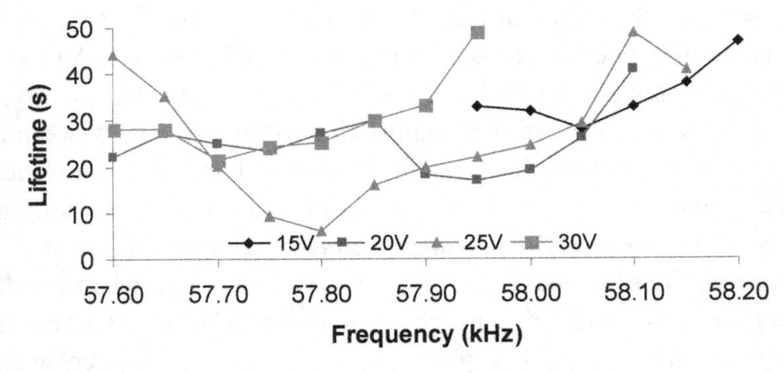

Fig.6.4.7. Lifetime vs operating frequency for 15 Vrms, 20 Vrms, 25 Vrms, and 30 Vrms.

The effect of vibration excitation location on the hypodermic needle was investigated. It was observed that when the vibration excitation location on the hypodermic needle was shifted away from point *P* (0, 6 mm, 0), the droplet was easier to be atomized. This is because the distance from point *P* to the tip is close to a quarter of the wavelength of water in the hypodermic needle, and the vibration excitation becomes stronger when the excitation location is shifted from this point. It was also observed that the hypodermic needle diameter had effect on the droplet size and rotation. Fig. 6.4.8 shows the maximum diameter of droplet versus vibration excitation location for three different types of hypodermic needles (blunt type), i.e. 25G, 30G and 33G needles. 25G, 30G and 33G hypodermic needles have approximately identical length (=10.5 mm) but different inner and outer diameters. The inner diameter of 25G, 30G and 33G hypodermic needle are 353μm, 172 μm and 152 μm, respectively; the outer diameter of 25G, 30G and 33G hypodermic needle are 540 μm, 311 μm and 252 μm, respectively. In the experiment, the operating voltage and frequency are 20 Vrms and 58.87 kHz, respectively. The 25G hypodermic needle generates larger droplets because it has larger inner and outer diameter, and the capillary effect has less effect on the water inside. According to observation, it is not easy to rotate the droplet in this case. This may be caused by the larger size of the droplets. The 30G and 33G hypodermic needles have smaller

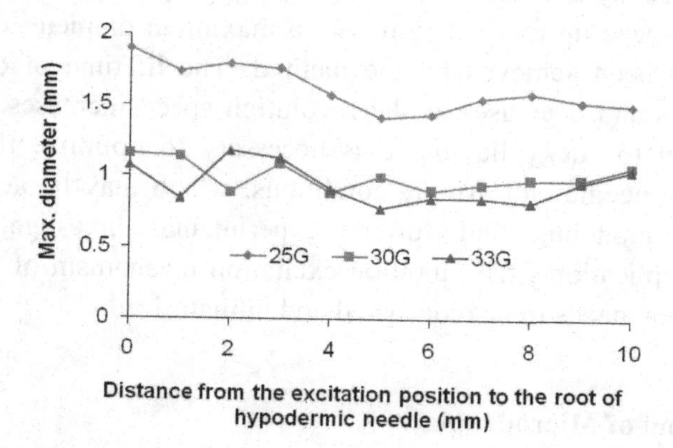

Fig. 6.4.8. Maximum droplet diameter versus vibration excitation position for different hypodermic needles.

droplets because they have smaller diameters and the capillary effect has larger effect on the water inside. According to observation, the droplets are easier to be atomized and not stable when 33G needle is used. This may result from a larger sound intensity in the water inside, which is caused by its thinner inner diameter.

Liquid such as cooking oil (viscosity = 50~100 centipoise), which has much larger viscosity than water (viscosity = 1 centipoise) was also tested. It was difficult to form the droplet at the tip by the ultrasonic transducer shown in Fig.6.4.1. This is because ultrasound in the cooking oil is weaker than that in the water due to the larger viscosity of cooking oil, which causes a larger acoustic damping. The surface tension of cooking oil ($\approx 32\times10^{-3}$ N/m) is about half of that of water ($\approx 72 \times 10^{-3}$ N/m). So the surface tension of liquid is not the determining factor that causes the failure in the generation of oil droplet. It is expected that a larger vibration or a stronger ultrasonic transducer is needed to form the droplets of sticky liquid.

6.4.3 *Summary*

We have shown that a hollow needle mechanically driven by an acoustic needle is capable to generate and rotate a small water droplet at its tip. The revolution speed, maximum diameter, and lifetime of the droplet can be controlled by the transducer's operating frequency and voltage. A revolution speed up to 1002 rpm with a maximum diameter of 4.3 mm droplet has been achieved by the method. The lifetime of droplets is several 10 s and decreases as the revolution speed increases. To apply this method to sticky liquids, it is necessary to optimize the size of hypodermic needle and driving conditions, which may be achieved by theoretical modeling and further experimental investigations. For practical applications, the vibration excitation mechanism of the device in this section needs to be redesigned and miniaturized.

6.5 Merging of Microdroplets

This section demonstrates a method to merge microdroplets by a wirelessly driven piezoelectric stage [10]. The contents of this section include the experimental setup, principle, results and discussion of this method.

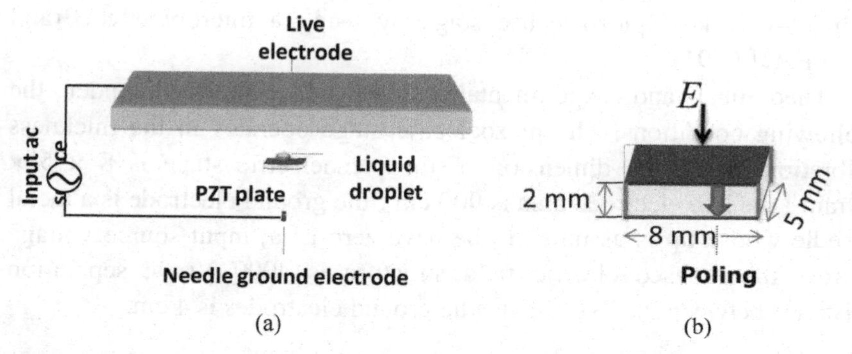

Fig. 6.5.1. (a) Experimental setup for the merging of droplets by a piezoelectric stage wirelessly driven by focused electric field. (b) Configuration of a wirelessly driven piezoelectric stage operating in the thickness vibration mode.

6.5.1 *Experimental setup and operating mechanism*

Figure 6.5.1 (a) shows the experimental setup for the merging of microdroplets by a piezoelectric stage wirelessly driven by focused electric field. With a square brass plate used as a live electrode, a stainless steel needle ground electrode is used to focus the ac electric field to enhance the electric energy transmission to piezoelectric stage. The needle ground electrode is placed below perpendicular to the live electrode which is suspended above the piezoelectric stage. The piezoelectric stage is placed equidistantly in between the live and needle ground electrodes of the focused electric field structure. The live electrode area is 30 × 30 cm^2. Fig. 6.5.1(b) shows the configuration of the wirelessly driven piezoelectric stage operating in the thickness vibration mode. The piezoelectric stage is made of lead zirconate titanate (PZT) ceramic material (Fuji C-203). It is poled along the thickness direction. Piezoelectric charge constant d_{33}, mechanical *Q*, dissipation factor *tanδ*, and relative dielectric constant $\varepsilon_{33}^{T}/\varepsilon_0$ are 325×10^{-12} m/V, 2000, 0.3 and 1450, respectively. The stage has silver electrode on its top and bottom surfaces. The dimension of the piezoelectric stage used in the experiments is $8 \times 5 \times 2$ mm^3. The vibration direction and applied electric field both are parallel to the poling direction. Droplets of volume varying from 0.5 μl to 2 μl are dispensed on to the top surface of wirelessly driven piezoelectric stage by using a micropipette (Brand CAPP, C1 9291).

Theoretical and experimental studies are performed under the following conditions. The piezoelectric stage operates in the thickness vibration mode; the dimension of the piezoelectric stage is $8 \times 5 \times 2$ mm^3; the live electrode area is 900 cm^2; the ground electrode is a metal needle whose tip is assumed to be have zero area; input source voltage across the focused electric field structure is 4000 Vrms; separation distance between the live and needle ground electrodes is 4 cm.

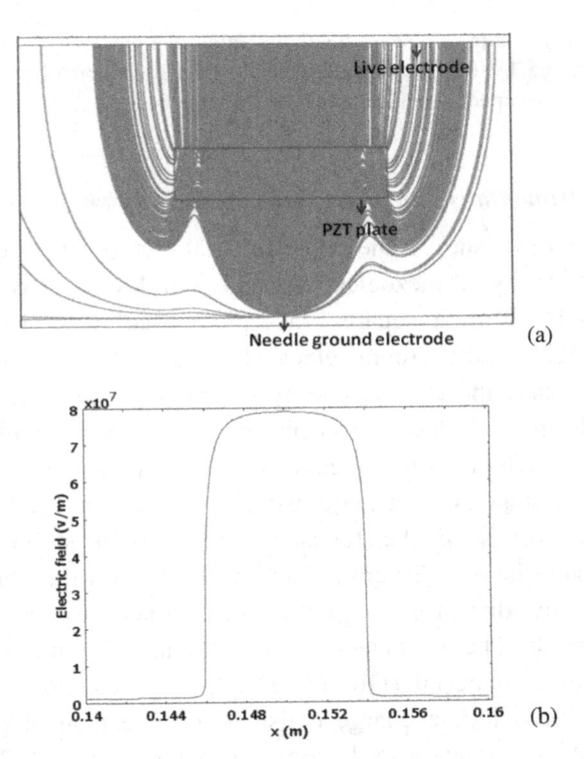

Fig. 6.5.2. (a) Calculated 2D electric field pattern around the piezoelectric stage wirelessly driven by focused electric field. (b) Distribution of electric field on the surface of wirelessly driven piezoelectric stage along the x-direction.

When an ac electric field produced from the focused electric field generator penetrates the piezoelectric stage, a mechanical vibration can be stimulated in the stage by the converse piezoelectric effect. When the frequency of ac electric field is close to mechanical resonance frequency of the piezoelectric stage, a strong enough mechanical resonance vibration can be excited in the stage. This mechanical vibration transmitted to the microdroplets placed on the surface of wirelessly driven piezoelectric stage, reduces the intermolecular forces of attraction or Vanderwaal's force between the water molecules. The microdroplets can go closer to each other with the vibration of the piezoelectric stage and merge after some time period.

6.5.2 *Results and discussion*

The finite element method (COMSOL Multiphysics) simulation has been carried out in order to assess the electric field on the surface of the

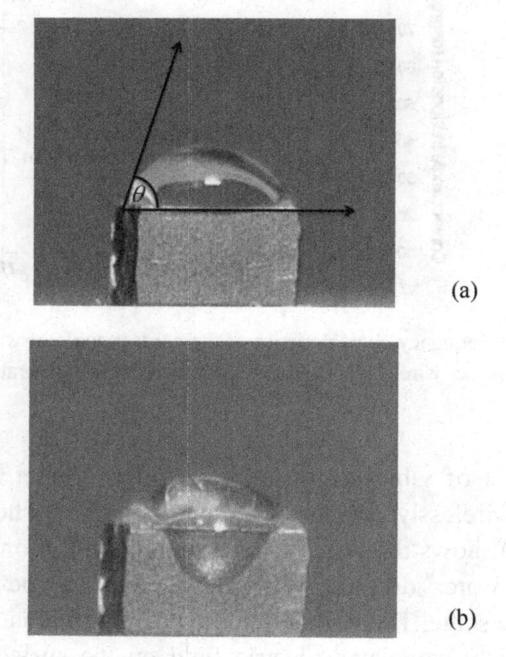

Fig. 6.5.3. (a) Water droplet on top surface of the wirelessly driven piezoelectric stage without vibration. (b) Water droplet flows down due to the vibration of piezoelectric stage.

piezoelectric stage without droplets on its top surface. Fig. 6.5.2(a) shows the calculated 2-D electric field pattern around the piezoelectric stage wirelessly driven by focused electric field. The distribution of electric field on the surface of the piezoelectric stage along the x-direction is shown in Fig. 6.5.2(b). It is seen that the electric field on the surface of the piezoelectric stage is non-uniform. It is found that the calculated average value of the electric field on the surface of the piezoelectric stage is 7.2×10^7 V/m for a piezoelectric stage with the dimensions of $8 \times 5 \times 2$ mm^3, optimum live electrode area of 900 cm^2, 4 cm live and needle ground electrodes separation, and an input source voltage of 4000 Vrms across the live and needle ground electrodes of focused electric field structure. The vibration displacement of the wirelessly driven piezoelectric stage is measured to be 0.1 μm (rms value).

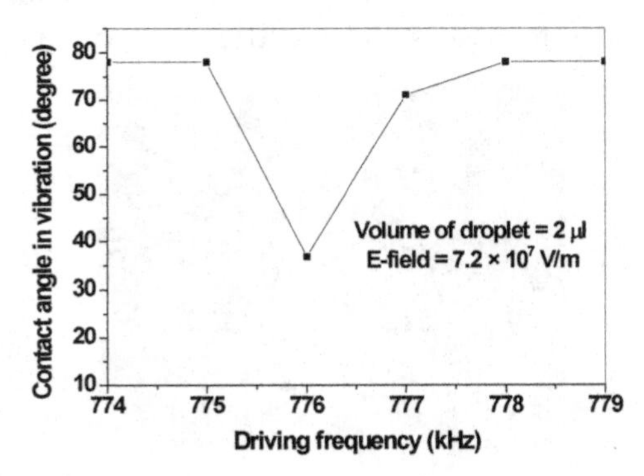

Fig. 6.5.4. The frequency characteristics of the contact angle of water droplet placed on the surface of the wirelessly driven piezoelectric stage operating in the thickness vibration.

The effect of vibration on a single microdroplet placed on the top surface of wirelessly driven piezoelectric stage is shown in Fig. 6.5.3. Fig. 6.5.3(a) shows the water droplet without vibration, and Fig. 6.5.3(b) shows the water droplet flows down due to the vibration of the piezoelectric stage. The volume of the water droplet is 2 μl, and the value of theoretically calculated electric field on the surface of piezoelectric stage is 7.2×10^7 V/m. It is observed that at resonance frequency of 776 kHz, the water droplet flows down from top surface of the piezoelectric

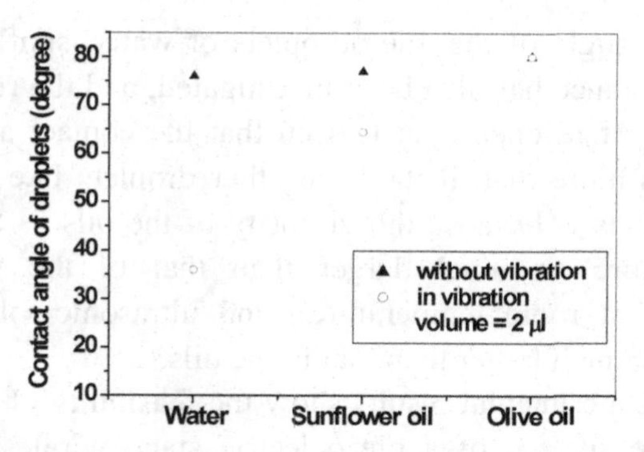

Fig. 6.5.5. The contact angle of liquid droplets on the horizontal surface of the wirelessly driven piezoelectric stage without vibration and in vibration.

stage as the vibration of the stage attains the maximum. The water droplet starts to flow down about 24 seconds after the vibration is excited. If the wirelessly driven piezoelectric stage is detuned from resonance, there is no flow-down of the water droplet. In order to illustrate the influence of electric field on the water droplet, the same experiment is carried out on the surface of a plastic stage, which has no vibration. It is observed that there is no change in the shape of water droplet after 24 seconds, and no flow-down also even after 15 minutes of experimentation. From this experiment, it is confirmed that the water droplets cannot merge without the ultrasonic vibration.

The frequency characteristics of the contact angle of water droplet placed on the surface of the wirelessly driven piezoelectric stage is shown in Fig. 6.5.4. It is seen that the contact angle θ made by the tangent line to the water droplet at the edge and the surface of the piezoelectric stage decreases at resonance frequency. It is observed that at resonance frequency of 776 kHz, the contact angle of water droplet in vibration reduces from 76.45° to 38.12°. This is because at resonance the vibration of the piezoelectric stage is maximum. When the vibration of the piezoelectric stage goes higher, the intermolecular force of attraction between the water molecules is reduced, and the water droplet tends to flatten. Thus, the contact angle measured is minimum at the maximum vibration. This experimental result excluded the possibility that the electric field in the droplet directly causes the intermolecular force change.

The contact angle of the microdroplets of water, sunflower oil, and olive oil at resonance has also been investigated, and the result is shown in Fig. 6.5.5. At resonance, it is seen that the contact angle of water droplet reduces more than that of the other droplets like sunflower oil and olive oil. This is because the viscosity of the oils (~ 80 mPa · S at room temperature) is much larger than that of the water droplet (0.894 mPa · S at room temperature), and ultrasonic vibration in the water droplet is much larger than that in the oils.

The above experimental results show the feasibility of microdroplets merging on the surface of a piezoelectric stage wirelessly driven by electric field. Fig. 6.5.6 shows the merging of two water microdroplets placed with a fixed separation distance on the surface of the wirelessly driven piezoelectric stage. It is observed that at zero second of vibration, the two water droplets of volume 0.8 μl are away from each other at a distance of 0.6 mm, shown in Fig. 6.5.6(a). After 10 seconds of sonication by the piezoelectric stage, the two microdroplets flattened and started approaching towards each other, as depicted in Fig. 6.5.6(b). The two water microdroplets become closer as the sonication time increases, and then suddenly merged 16 seconds after the excitation of piezoelectric stage, which is shown in Fig. 6.5.6 (c). The mechanical vibration of the piezoelectric stage transmitted to the microdroplets can cause the decrease in intermolecular force of attraction in the microdroplets which cause the microdroplets to flatten and merge.

(a) (b) (c)

Fig. 6.5.6. Merging of two water droplets placed with a fixed separation distance on the surface of a wirelessly driven piezoelectric stage operating in the thickness vibration mode. (a) without vibration; (b) after 10 seconds of sonication ; (c) after 16 seconds of sonication.

Experimentally it has been observed that the merging speed of microdroplets depends on the vibration displacement at resonance of piezoelectric stage, separation distance and volume of microdroplets. The dependence of measured time for merging of water microdroplets on the separation distance and volume of the water droplets is shown in Fig. 6.5.7. It has been found that the time for the merging decreases with the decrease of the separation distance between the two microdroplets, and the increase of the volume of each droplet. At resonance frequency of 776 kHz, two water droplets each of volume 0.8 μl separated by a distance of 0.2 mm merge in 8 seconds after the vibration of 0.1 μm (rms) is excited in the wirelessly driven piezoelectric stage. It is also seen that the merging speed of larger volume of microdroplets is significantly higher than that of smaller volume microdroplets. Once the microdroplets start to move, the microdroplet with larger volume flows faster because of its larger inertia. It is observed that before vibration of the piezoelectric stage the measured temperature of water droplets was 25.6 °C whereas during the merging 25.8 °C. Thus, the temperature rise of the microdroplets is very small, and its effect on the merging may be ignored.

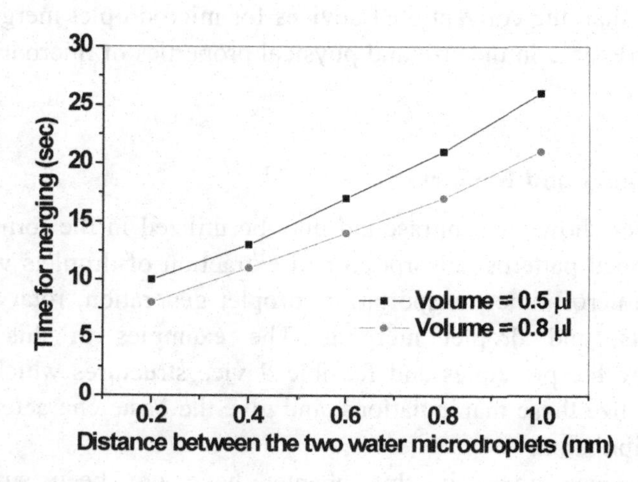

Fig. 6.5.7. Dependence of the time for merging of water droplets on the distance and volume of water droplets.

6.5.3 *Summary*

A method to merge microdroplets by using a wirelessly driven piezoelectric stage is demonstrated. The separated microdroplets (volume varying from 0.5 μl to 2 μl) to merge are dispensed onto the top surface of a piezoelectric stage. The piezoelectric stage is wirelessly driven by a focused electric field. The ultrasonic vibration of the piezoelectric stage is transmitted into the microdroplets and induces the merging of microdroplets. Experimentally, it has been observed that the time for merging of water microdroplets depends on the vibration displacement of piezoelectric stage, and separation distance and volume of microdroplets. The time for merging of two water microdroplets decreases with the decrease of separation distance between the two microdroplets, and the increase in the volume of microdroplets. It is seen that the merging speed of larger volume of microdroplets is significantly higher than that of smaller volume ones. At resonance frequency of 776 kHz, two water microdroplets each of volume 0.8 μl separated by a distance of 0.6 mm are merged together 16 seconds after a mechanical vibration displacement of 0.1 μm is excited in the piezoelectric stage with area of 40 mm^2. The proposed device has simpler structure and the potential to be smaller than the conventional devices for microdroplet merging. It is also more flexible in the size and physical properties of microdroplets to merge.

6.6 Summary and Remarks

This chapter shows that ultrasound may be utilized in the formation of acoustic lobed patterns, adsorption and extraction of droplets without a chamber, microfluidic transportation, droplet generation, rotary driving of droplets, and droplet merging. The examples in this chapter demonstrate the principles and feasible device structures which can be used to realize these manipulations, and give the basic characteristics of these manipulations.

The devices given in this chapter have not been sufficiently optimized. To fully optimize these devices, modeling them mathematically is the first step. This modeling involves applications of molecular dynamics, which will be a new branch of acoustics.

References

1. Hu, J. H., Zhu, H., Li, N. and Zhao, C. S. (2011). Sound induced lobed pattern in aqueous suspension film of micro particles, Sens. Actuators, A:Physical, 167 (1), pp. 77–83.
2. Hu, J. H., Li, N. and Zhou, J. J. (2011). Controlled adsorption of droplets onto anti–nodes of an ultrasonically vibrating needle, J. Appl. Phys., 110 (5), 054901.
3. Hu, J. H., Tan, C. L. and Hu, W. Y. (2007). Ultrasonic microfluidic transportation based on a twisted bundle of thin metal wires, Sens. Actuators, A, 135 (2), pp. 811–817.
4. Tan, Z. W., Teo, S. G. G. and Hu, J. H. (2008). Ultrasonic generation and rotation of a small droplet at the tip of a hypodermic needle, J. Appl. Phys., 104 (10), pp. 104902.
5. Hu, J. H., Nakamura, K. and Ueha, S. (1996). Optimum operation conditions of an ultrasonic motor driving fluid directly, Jpn. J. Appl. Phys, 35, pp. 3289–3294.
6. Hu, J. H., Cha, K. C. and Lim, K. C. (2004). New type of linear ultrasonic actuator based on a plate–shaped vibrator with triangular grooves, IEEE Transactions on Ultrasonics, Ferroelectrics, and Frequency Control, 55(10), pp. 1206–1208.
7. Hu, J. H., Li, G. R., Chan, H. L. W, and Choy, C. L. (2001). A standing wave–type noncontact linear ultrasonic motor, IEEE Transactions on Ultrasonics, Ferroelectrics, and Frequency Control, 48 (3), pp. 699–708.
8. Abramov, O. V. (1998) *High–Intensity Ultrasonics: Theory and Industrial Applications* (Gordon and Breach Scientific Publishers, Singapore) pp. 297.
9. Hu, J. H., Nakamura, K. and Ueha, S. (1999). A noncontact ultrasonic motor with the rotor levitated by axial acoustic viscous force, Electrochem. Commun. in Japan (Part III), 82 (4), pp. 56–63.
10. Bhuyan, S. (2010) *Wireless drive of Piezoelectric Components*, Chapter 7, (PhD thesis, Nanyang Technological University), pp. 139.

Chapter 7

Concluding Remarks

The application of ultrasound can be classified into two types. One is the information application such as the ultrasonic imaging and nondestructive test, and another is the actuation application. Systematic researches on the actuation application of ultrasound started from 30 years ago, and the ultrasonic motor is one of the main representatives of ultrasonic actuators. Ultrasonic manipulation technology provides various manipulation functions for micro/nano solids, soft materials, droplets and microfluid. It is a new development and branch of ultrasonic actuation technology. It can also be viewed as a new academic field, generated by the merge of physical acoustics, mechanical design, piezoelectrics and electrical equipment.

The functions of ultrasonic micro/nano manipulation, which have been realized experimentally, include

(1) contact and noncontact type trapping and transfer of micro solids;
(2) concentration and separation of micro solids;
(3) extraction of micro solids;
(4) spin and revolution of micro solids;
(5) removal of micro solids;
(6) contact and noncontact type trapping and transfer of single and multiple nanowires;
(7) controlled rotary driving of single nanowires;
(8) concentration and radial alignment of nano entities;
(9) adsorption and extraction of droplets without a chamber;
(10) microfluidic transportation;
(11) droplet generation;
(12) rotary driving of single droplets;

(13) merge of droplets, etc.

The merits of ultrasonic micro/nano manipulation include little selectivity to material property of manipulated samples, diversity in manipulation functions, compact device structures, devices easy to fabricate, and very low temperature rise at the manipulation location in some methods. The materials which have been successfully manipulated by ultrasound include metal, silicon, oxidation, cell, plant seeds, animal eggs, salt, sugar, flour, pollen, water, oil, etc. Ultrasonic micro/nano manipulation technology can not only handle single and multiple entities in a small manipulation area, but also remove dust particles in a relatively large area. So far, ultrasonic micro/nano manipulation has been using piezoelectric devices to generate the necessary vibration and ultrasonic field; thus the devices have a compact structure and easy to fabricate. In the nano manipulation using acoustic streaming, the temperature rise at the manipulation location can be lower than 0.1℃. The merits of ultrasonic micro/nano manipulation make it competitive and promising in the emerging market of micro/nano manipulation.

Due to the use of mechanical waves and vibration, a major demerit of ultrasonic micro/nano manipulation technology is that the methods based on acoustic radiation force and acoustic streaming cannot be used to directly manipulate the samples in vacuum, because there are no acoustic radiation force and acoustic streaming in vacuum. The traveling wave induced frictional driving and Chladni effect may be used to trap micro/nano entities in vacuum in a plane. But a 3D trapping of them in vacuum is still a challenge.

To develop the ultrasonic micro/nano manipulation technology, there are still lots of challenges. Some of them are listed as follows:

(1) to completely understand the behavior of nano entities in the physical effects of ultrasound described in Chapter 2;
(2) to effectively model the physical effects such as the acoustic streaming, frictional driving, sound induced intermolecular force change and so on, to understand the quantitative effects of working parameters such as the operating frequency, vibration strength, shape of the radiation sources, properties of acoustic medium, topology of acoustic field, etc., on these physical effects;

(3) to build a theory to answer the question that what kind ultrasonic field is the best for a particular manipulation and sample;
(4) to optimize structural topologies for the vibration excitation and manipulation in the manipulating devices to make them more compact and efficient;
(5) to increase the manipulation yield;
(6) to develop simple and low-cost manipulating systems for practical applications;
(7) to observe smaller objects.

Index

Printed in the United States
By Bookmasters